AF861376

STELLINGEN

Behorende bij het proefschrift 'Dynamical response of saturated and dry soils' door P. Hölscher

1 De analytische oplossing van Van der Grinten e.a. is een beter instrument om numerieke modellen voor twee-fasen dynamica te testen dan de analytische oplossing van Simon e.a., omdat de oplossing van Van der Grinten e.a. geen dynamische compatibiliteit verondersteld.

Simon, B.R., Zienkiewicz, O.C., Paul, D.K.
An analytical solution for the transient response of saturated porous elastic solids
International Journal for Numerical and Analytical Methods in Geomechanics, 1984, Vol. 8, pp. 381-398

Van der Grinten, J.G.M., Van Dongen, M.E.H., Van der Kogel, H.
Strain and pore pressure propagation in a water-saturated porous medium
Journal of Applied Physics, Vol. 62, No. 12, december, 1987, pp. 4682-4687

2 Bij reflectie van golven op een laagscheiding tussen een ondoorlatend materiaal en een twee-fasen materiaal wordt de spanning in het skelet beïnvloed door de golf van de tweede soort. Deze bevinding wijkt of van de conclusies van Geertsma en Smit.

Geertsma, J. & Smit, D.C.
Some aspects of elastic wave propagation in fluid-saturated porous solids
Geophysics, Vol. 26, No. 2, April 1961, pp. 169-181

3 Tijdens quasi-statische belasting van een funderingspaal wordt de grond rondom de paal dynamisch belast. De daarbij optredende waterspanningen en traagheidskrachten mogen in slappe bodem niet verwaarloosd worden bij de schatting van het statisch draagvermogen uit de gemeten quasi-statische weerstand.

4 De spanningen die ontstaan door een golfklap belasting op een golfbreker, kunnen nadelige invloed hebben op de geotechnische stabiliteit van de golfbreker. Dit effekt kan niet beoordeeld worden door alleen de waterbeweging in de golfbreker te beschouwen.

5 Bij dynamische verschijnselen waarbij de belastingstijd en de consolidatie tijd van dezelfde orde van grootte zijn en waarbij niet-lineair grondgedrag een rol speelt, is de toepassing van twee-fasen dynamica noodzakelijk.

6 Een foto-elastische proef, zoals die in paragraaf 5.4 van dit proefschrift beschreven is, leent zich uitstekend voor het verkrijgen van inzicht in de relatie tussen het statisch en dynamisch gedrag van de grond rondom de paalpunt.

7 Voor de berekening van trillingsuitbreiding langs het oppervlak van een elastische halfruimte zijn oplossingen die gebaseerd zijn op de aanname dat er uitsluitend vertikale vervormingen optreden, zoals deze aangegeven zijn door Barends en Verruijt, niet bruikbaar.

Barends, F.B.J.
Dynamics of elastic plates on a flexible subsoil
LGM-mededelingen, Part XXI, no. 2, June 1980

Verruijt, A.
Soil dynamics, college dictaat b28, TU-Delft, 1993

8 De tijdsintegrator volgens Lax is stabiel voor ongedempte systemen, maar niet voor zwaar gedempte systemen.

Hölscher, P., Barends, F.B.J
Finite difference scheme for wave transmission in a rubblemound breakwater,
Int. Jnl. Num. Meth. Eng., pp. 1129-1146, 1990

9 Voor een gelaagde bodem is de exponent n in de gegeneraliseerde formule van Barkan sterk frequentie afhankelijk.

Barkan, D.D.,
Dynamics of soils and foundations, McGraw-Hill Book Compagny, 1962

10 Het democratische gehalte van een parlementaire democratie wordt sterk verhoogd, als het mogelijk is een (protest) stem op een 'lege zetel' uit te brengen.

Bij overschrijding van de kiesdrempel door het totale aantal stemmen op deze 'lege zetel', leiden deze stemmen tot een evenredig aantal lege zetels in het gekozen orgaan. Bij besluitvorming worden 'lege zetels' beschouwd als afwezige leden en zijn besluiten slechts aangenomen als de helft van het quorum het voorstel steunt.

11 De touristische waarde van reizen met dubbeldeks treinen neemt toe met het aantal geluidsschermen langs de spoorlijn.

12 Paardesport wordt ruitersport als ruiter en paard van plaats wisselen.

Dynamical response of saturated and dry soils

Dynamical response of saturated and dry soils

(Dynamische reactie van verzadigde en droge grond)

PROEFSCHRIFT

ter verkrijging van de graad van doctor
aan de Technische Universiteit Delft,
op gezag van de Rector Magnificus prof. ir. K.F. Wakker,
in het openbaar te verdedigen ten overstaan van een commissie,
door het College van Dekanen aangewezen,
op maandag 6 februari 1995 te 16.00 uur
door

Paul HÖLSCHER

civiel ingenieur
geboren te Velsen

Dit proefschrift is goedgekeurd door de promotor
Prof. dr. ir. F.B.J. Barends

Samenstelling promotie commissie:

Rector Magnificus
Prof. dr. ir. F.B.J. Barends
Prof. dr. ir. A. Verruijt
Prof. dr. J.D. Nieuwenhuis
Prof. ir. A.F. van Tol
Prof. dr. ir. W. van Impe
Prof. ir. A.C.W.M. Vrouwenvelder
Dr. Ir. M.E.H. van Dongen

Published and distributed by:

Delft University Press
Stevinweg 1
2628 CN DELFT
the Netherlands

telephone +31 (0)15 783254
fax +31 (0)15 781661

CIP-GEGEVENS KONINKLIJKE BIBLIOTHEEK, DEN HAAG

Hölscher, Paul

Dynamical response of saturated and dry soils / Paul
Hölscher. - Delft : Delft University Press. - Ill.
Proefschrift Technische Universiteit Delft. - Met lit.
opg.
ISBN 90-407-1073-2
Trefw.: poreuze media ; dynamica.

Aan Rinske

Acknowledgement

The work reported in this thesis is supported by the Dutch Technology Foundation STW and Delft Geotechnics.

The support and the hospitality of the Geotechnical Laboratory of the department of Civil Engineering of Delft University of Technology are kindly acknowledged.

Furthermore, I would like to thank:
my promotor, prof. Frans Barends, for his continuous encouragement;
dr. Hendrikus Allersma, for the useful suggestions for the experiments;
mr. Guus Segal, for the assistance using the finite element package Sepran;
mr. Ab Mensinga, mr. Joop van Leeuwen and mr. Han de Visser, for their assistance during the research;
prof. F. Oka and prof. A. Yashima for the possibility to use the non-linear hardening model in this research;
dr. Hans Teunisse for the useful discussion over finite element modelling and non-linear models;
The members of the 'gebruikerscommissie' (mrs. C.N.M. Janzs, prof. A.C.W.M. Vrouwenvelder, dr. M.E.H. van Dongen, dr. H.G.B. Allersma, mr. C.E.H.M. Buys, mr. A. van der Toorn, mr. J. Lindenberg) for their questions and suggestions.

SUMMARY

Dynamical response of saturated and dry soils

The research described in this thesis is concerned with the behaviour of soil under dynamical loading where the rates of loading and consolidation are of the same order, and are sufficiently high that allowance has to be made for inertia effects.

The first step involves an understanding and description of the behaviour of saturated material under dynamical loading. The second step is an answer to the question whether a two-phase description is necessary for practical problems, or whether a simpler single phase description would provide sufficient accuracy.

The study begins with a description of the dynamical behaviour of saturated porous media, as proposed by Biot in 1956. Special attention is paid to the interaction between the solid and the fluid, and to the conditions at the interface between different materials.

The reflection and transmission of waves near interfaces are studied analytically for one-dimensional geometry (i.e. waves propagated perpendicular to the interface). The solution is calculated in the frequency-domain. Time-domain solutions are calculated using fast Fourier transform.

For one-dimensional shear wave propagation in a linear elastic material, the two-phase behaviour is not relevant for many practical problems in soil mechanics. For compression wave propagation, the solid stress near the interface can be significantly affected. At the interface, a reduction of solid stress up to 30% (of the value obtained by single phase analysis) is encountered for practical values of soil properties. The Rayleigh wave at the surface of a saturated porous medium is hardly affected by the two-phase behaviour.

To solve two-dimensional problems with a complex geometry and a complex material behaviour, a finite element method is developed, the program 'Titan'. It links the theories of wave propagation and consolidation over the entire time-domain. In Titan, a non-linear model for the solid is implemented.

Experimental insight into the behaviour of a saturated porous medium is obtained by three dynamical experiments. One of these experiments is simulated with the program Titan.

The first experiment consists of a pile-driving simulator suitable for use in a laboratory. The differences in the wave propagation in saturated and dry sand under the toe of a dynamically-loaded pile are demonstrated.

The second experiment is a full-scale pile-driving test. The pore-water pressure and the acceleration in the soil close to the pile toe are measured during driving, dynamic testing and quasi-static testing. A complete velocity field around the pile toe is derived.

The third experiment shows the possibility of measuring by a photoelastic method the dynamical solid stress in an oil-saturated granular material.

Finally, the application of Titan to two practical problems is shown.

The first problem is concerned with the simulation of the pile-driving simulator. The experimental results are confirmed for dense sand.

The second problem relates to the effects of a wave attack on a partially saturated breakwater. High stresses are found at the interface between the armour and the core. The importance of using the two-phase approach is demonstrated.

SAMENVATTING

Dynamische reactie van verzadigde en droge grond

Het onderzoek, dat wordt beschreven in dit proefschrift, behandelt het gedrag van grond onder dynamische belasting, waarbij de belastingtijd en de consolidatietijd in orde van grootte gelijk zijn en zo hoog dat traagheidseffecten niet verwaarloosd mogen worden.

De eerste stap omvat het begrijpen en beschrijven van het gedrag van verzadigde poreuze media onder dynamische belasting. De tweede stap bestaat uit het antwoord op de vraag of de beschrijving met twee fasen nodig is voor praktische problemen, of dat de eenvoudigere één-fase beschrijving voldoende nauwkeurig is.

Er is uitgegaan van de beschrijving van het dynamische gedrag van verzadigde grond, zoals dit door Biot in 1956 gemodelleerd is. Hierbij is speciale aandacht besteed aan de interactie tussen het skelet en de porievloeistof en de voorwaarden die gelden op de laagscheidingen tussen twee verschillende materialen.

De reflectie en transmissie van golven ter plaatse van laagscheidingen is analytisch opgelost in het frequentie domien voor één-dimensionale geometrie, waarbij de voortplantingsrichting van de golven loodrecht op de laagscheiding staat. De oplossing in het tijdsdomein is verkregen met behulp van numerieke Fourier transformaties.

Voor de meeste praktische problemen in de grondmechanica is het twee-fasen effect niet relevant voor de voortplanting van schuifgolven in een één-dimensionaal, lineair elastische materiaal. Voor de voortplanting van compressie golven geldt dat de spanningen in het skelet nabij een laagscheiding wel significant beïnvloed worden. Voor praktische situaties is een afname tot 70% van de verre veld waarde gevonden.

De voortplanting van Rayleigh golven langs het oppervlak van een verzadigd poreus medium wordt nauwelijks beïnvloed door het twee-fasen gedrag.

Om problemen met een twee-dimensionale complexe geometrie en complexe materiaal eigenschappen op te kunnen lossen is een eindig elementen programma ontwikkeld, het programma Titan. Dit programma legt een relatie tussen de theorie van de golfvoortplanting en de theorie van consolidatie. Een model om het niet-lineaire gedrag van het skelet te beschrijven is in Titan geïmplementeerd.

Experimenteel inzicht in het dynamische gedrag van verzadigde poreuze media, is verkregen door het uitvoeren van drie experimenten. Voor één van deze experimenten is de berekening met het programma Titan beschreven.

Het eerste experiment bestaat uit een hei-simulator, die in een laboratorium werkt. Met behulp van deze hei-simulator is het verschil tussen golfuitstraling vanaf de paalpunt in droge en verzadigde grond getoond.

Het tweede experiment bestaat uit veld metingen tijdens heien. De dynamische waterspanningen en versnellingen in de grond nabij de paalpunt zijn in drie situaties gemeten: tijdens het heien, het dynamisch proefbelasten en het quasi-statisch proefbelasten. Uit deze metingen is een volledig snelheidsveld van de grond rondom de paalpunt afgeleid.

Het derde experiment toont aan dat het mogelijk is dynamische spanningen in het skelet te meten in een met olie verzadigd monster van grbroken glas door gebruik te maken van foto-elasticiteit.

Tenslotte is de toepasbaarheid van het programma Titan aan de hand van twee praktische problemen geïllustreerd.

Het eerste probleem betreft de simulatie van de proeven met de hei-simulator. Voor dicht gepakt zand worden resultaten gevonden die overeenkomen met de experimentele resultaten.

Het tweede probleem betreft de invloed van een golfklap op een gedeeltelijk met water verzadigde golfbreker. Hoge spanningen treden op ter plaatse van de laagscheiding tussen de afdeklaag en de kern. Het belang van de twee-fasen benadering wordt hiermee aangetoond.

TABLE OF CONTENTS

1 INTRODUCTION

1.1 Background to the study

The present study, 'Dynamical response of saturated and dry soils', describes an attempt to combine in one model two disciplines of mechanics, namely the theory of consolidation and the theory of wave propagation.

Since the beginning of this century, engineers have been developing methods for describing the behaviour of soil which take into account the fact that, in principle, soil is a two-phase material, consisting of a solid (skeleton) and a fluid. Terzaghi [1923] developed the principles which led to the consolidation theory. The consolidation process, related to the strength, settlement and stability of structures, is one of the main time-dependent problems in soil-mechanics. Nowadays, the finite element method allows the simulation of soil behaviour for complex geometries, introducing realistic non-linear material behaviour.

The basic theory of wave propagation in single phase materials was developed at the beginning of this century. Due to the complexity of the problem, however, further development took place slowly. In the second half of this century, the development of the field of dynamics took place synchronously with the development of computational mechanics. Seismic and acoustic research are important current fields of application. Because, in general, the frequency in these applications is high and the amplitudes are small, linear theory is applicable.

For many practical dynamical problems in soil mechanics, the single phase approximation is sufficient, since the time-scales of the dynamic loading and consolidation are of different orders. If the rate of loading is high and the soil is not very permeable, the behaviour is modelled as fully undrained. If the rate of loading is sufficiently low, inertia forces can be ignored. Depending on the permeability and the rate of loading, the behaviour is modelled as drained or undrained.

The research described in this thesis deals with loading on soils where the rate of loading and the rate of consolidation are of the same order, and are sufficiently high that allowance has to be made for inertia effects. This study has two aims. First, the behaviour of saturated material under dynamical loading has to be understood and described. Second, an attempt is made to answer the question whether the complicated two-phase description is necessary for practical problems, or whether a simpler single phase description can be used.

Practical problems in this field are pile-driving, dynamical (or vibrational) densification, wave loading on dykes and breakwaters, loading of rail embankments by passing high-speed trains, and the behaviour of foundation piles loaded by rapidly rotating machinery. Because of these practical problems, the linearity of the solid behaviour cannot be assumed, in contrast to seismological research. For some problems, e.g. pile-driving and soil densification, failure and permanent deformation are the desired results of the process.

For other problems, (cumulation of) non-linear deformations may lead to undesired failure of soil structures. Assessment of realistic non-linear soil behaviour in a dynamic two-phase environment is a serious complication, which must undoubtedly be overcome if one is to acquire a better comprehension of practical problems.

1.2 Outline of the study

In the fifties, Biot [1956] extended Terzaghi's theory into one which describes the dynamical behaviour of saturated soil. Biot established the existence of two compressional waves, but it took more than 20 years to prove their existence experimentally [Plona, 1980, Van der Grinten et al., 1985]. Biot's theory has been the starting-point for many theoretical and practical studies in the field of dynamics.

Chapter 2 presents a summary of Biot's linear theory. Attention is paid to the interaction between the solid and the fluid. The conditions at the interface between different materials are described.

In chapter 3, analytical solutions are investigated. In general, the wave of the second kind is highly damped. It can therefore be ignored in far field solutions. This chapter demonstrates its influence on solutions close to interfaces between saturated porous materials, and between a saturated porous material and an impermeable material (near field solutions). Solutions in both the frequency- and time-domains are presented, based on the work of Van der Grinten et al. [1984]. At the end of this chapter, the propagation of harmonic Rayleigh waves in the far field is discussed.

In practice, the solution of the wave propagation on non-homogeneous two-dimensional domains with saturated porous materials cannot be found using analytical methods. Several descriptions are presented in the literature, e.g. Zienkiewicz and Bettes [1983] and Prevost [1985]. In chapter 4, a numerical method of solution based on the finite element method is developed. This method is new, and is applicable over a wide frequency range. In contrast to other methods, it links the theory of wave propagation to the theory of consolidation over the entire time-domain.

Chapter 4 also presents a practical non-linear model for the behaviour of the solid (skeleton) suitable for dynamical analysis. This model, originally proposed by Oka [1992] for describing the liquefaction of sand, is improved and extended for the description of soil behaviour during compression.

Three experiments relating to pile-driving are described in chapter 5. The aim of the first experiment is the experimental verification of the hypothesis that wave propagation under a dynamically loaded pile in a saturated soil is essentially different from wave propagation in a dry soil. A pile-driving simulator for use in the laboratory has been constructed. The second experiment consists of the measurement of the soil motion and the pore-water pressure in the field close to the toe of a pile during driving and dynamic testing. The third experiment shows the possibility of measuring by the photoelasticity method the dynamical solid stress in an oil-saturated granular material.

Finally, application of the numerical model developed in this study is demonstrated in chapter 6. The model is applied to one of the tests described in chapter 5, and to a practical problem of a wave attack on a rubble mound breakwater.

2 GENERAL DESCRIPTION OF DYNAMICS OF POROUS MEDIA

2.1 Overview

In this chapter the equations which describe the dynamical behaviour of a saturated porous medium are discussed. The interaction between the solid and the fluid are studied. The resulting equations are rewritten by introduction of the specific discharge as new field variable. Finally the initial conditions, the boundary conditions and the conditions at interfaces between different materials are derived.

2.2 Equations for porous material

2.2.1 General equations in three dimensions

The soil is described according to the theory of Biot [1956]. Biot supposed that

- the volumetric porosity is equal to the area porosity (statistically isotropic)
- the medium is homogeneous
- the wavelength is much longer than the particle size.

The kinematic equations, constitutive equations and dynamic (equilibrium) equations are presented. The equations are limited to infinitesimal displacements and linear constitutive models. Later non-linear constitutive models will be considered. The equations hold for a homogeneous field. The conditions at interfaces and boundaries are discussed in the following sections.

The index notation in cartesian coordinates is adopted. The summation convention for subscripts i,j and k is used.

The following equations are used as basic equations for this research

Kinematic equations

$$\begin{aligned} &\textit{solid:} \quad \varepsilon^s_{ij} = \frac{1}{2}\left(u^s_{i,j} + u^s_{j,i} \right) \\ &\textit{fluid:} \quad \varepsilon^f_{ii} = u^f_{i,i} + \frac{1-n}{n} u^s_{i,i} \end{aligned} \tag{2.1}$$

Constitutive equations

$$solid: \quad \sigma^s_{ij} = (K_s - \tfrac{2}{3}G_s)\,\varepsilon^s_{kk}\,\delta_{ij} + 2G_s\varepsilon^s_{ij}$$

$$or\ \sigma^s_{ij} = D_{ijkl}\,\varepsilon^s_{kl} = [\,(K_s - \tfrac{2}{3}G_s)\,\delta_{ij}\,\delta_{kl} + G_s(\delta_{ik}\,\delta_{jl} + \delta_{il}\,\delta_{jk})\,]\,\varepsilon^s_{kl} \tag{2.2}$$

$$fluid: \quad \pi = K_f\varepsilon^f_{kk}$$

$$interaction: \quad R_j = b\,(\dot{u}^f_j - \dot{u}^s_j)$$

Dynamic equations

$$\sigma^s_{ij,i} + (1-n)\pi_{,j} + R_j = [(1-n)\rho_s + \rho_a]\,\ddot{u}^s_j - \rho_a\ddot{u}^f_j$$
$$n\pi_{,j} - R_j = -\rho_a\ddot{u}^s_j + [\rho_a + n\rho_f]\,\ddot{u}^f_j \tag{2.3}$$

with
- u — displacement (m)
- ε — strain (-)
- σ^s — effective stress (N/m^2), positive for tension
- π — fluid stress (N/m^2), positive for tension (= 'minus' pore pressure)
- ρ — volumetric mass (kg/m^3)
- δ_{ij} — Kronecker delta (-)
- n — effective porosity (-)
- K_s — bulk modulus of (drained) solid (N/m^2)
- G_s — shear modulus of solid (N/m^2)
- D_{ijkl} — elastic stiffness tensor
- K_f — bulk modulus of fluid (N/m^2)
- R_j — volumetric friction of fluid on solid (N/m^3)

$$R_j = b(\dot{u}^f_j - \dot{u}^s_j) \tag{2.4}$$

with b friction coefficient, for Darcy flow $b = n^2\mu/K$
- μ: dynamic viscosity (kg/sm)
- K: intrinsic permeability (m^2)

The super- and subscripts s and f refer to the solid and the fluid respectively. The apparent volumetric mass ρ_a will be discussed in the next section. The subscript comma ',' refers to partial derivative with respect to the space dimension, the superscript dot '˙' refers to the partial derivative with respect to time.

2.2.2 Discussion on interaction terms

Three interactions between the two phases occur:

- one depending on the (relative) volume changes of the two phases (here called the kinematic interaction)
- one depending on the relative velocity of the two phases (generally known as friction or viscous interaction)
- one depending on the relative acceleration of the two phases (here called the dynamic interaction)

The kinematic equation for the fluid reads

$$\varepsilon^f_{kk} = u^f_{k,k} + \frac{1-n}{n} u^s_{k,k} \tag{2.5}$$

The second term of the right hand side is the kinematic interaction term. It shows that a volumetric strain of the solid leads to a volumetric strain of the fluid.

The need for this term can be explained by some theoretical experiments on a small element of saturated porous material. The volume of the element is denoted by $V = \Delta x_1 \Delta x_2 \Delta x_3$. The pores are filled by the fluid, whose volume is nV. The remaining part of the volume V is occupied by the solid, so the volume of the solid equals (1-n)V. In the experiments the displacements of the two phases are prescribed.

In the first experiment the displacement u^f of the fluid is non-zero and the displacement in the solid u^s is zero. Now the strain in the solid and the strain in the fluid are calculated

$$\varepsilon^s_{kk} = u^s_{k,k} = 0$$

$$\begin{aligned} \varepsilon^f_{kk} &= \Delta V^f / V^f = \\ &\frac{(n\Delta u^f_1 \Delta x_2 \Delta x_3 + n\Delta u^f_2 \Delta x_1 \Delta x_3 + n\Delta u^f_3 \Delta x_1 \Delta x_2)}{n\Delta x_1 \Delta x_2 \Delta x_3} \\ &= u^f_{1,1} + u^f_{2,2} + u^f_{3,3} = u^f_{k,k} \end{aligned} \tag{2.6}$$

So Equation (2.5) is satisfied.

In the second experiment the displacement u^s of the solid is non-zero and the displacement in the fluid u^f is zero. The strain in the solid and the fluid are calculated again.

Due to the strain in the solid, the porosity of the solid is changed. This change in porosity leads to a change in the volume which is occupied by the fluid. However, the fluid does not displace (flow), so the fluid which occupied the pores before the experiment must expand. The expansion equals the volume change of the skeleton due to the displacement u^s. This reasoning leads to the equations

$$\varepsilon_{kk}^{s} = \frac{\Delta V^{s}}{V^{s}}$$

$$= \frac{(1-n)\Delta u_1^{s}\Delta x_2\Delta x_3 + (1-n)\Delta u_2^{s}\Delta x_1\Delta x_3 + (1-1)\Delta u_3^{s}\Delta x_1\Delta x_2}{(1-n)\Delta x_1\Delta x_2\Delta x_3}$$

$$= u_{1,1}^{s} + u_{2,2}^{s} + u_{3,3}^{s} = u_{k,k}^{s}$$

$$\Delta V^{s} = \Delta V^{f} \Rightarrow \tag{2.7}$$

$$\varepsilon_{kk}^{f} = \frac{\Delta V^{f}}{V^{f}} = \frac{\Delta V^{s}}{V^{f}} =$$

$$\frac{(1-n)\Delta u_1^{s}\Delta x_2\Delta x_3 + (1-n)\Delta u_2^{s}\Delta x_1\Delta x_3 + (1-n)\Delta u_3^{s}\Delta x_1\Delta x_2}{n\Delta x_1\Delta x_2\Delta x_3}$$

$$= \frac{(1-n)}{n}(u_{1,1}^{s} + u_{2,2}^{s} + u_{3,3}^{s}) = \frac{(1-n)}{n}u_{k,k}^{s}$$

So, Equation (2.5) is satisfied.

The third and last experiment refers to the situation in which the displacements of both phases are non-zero. This is a combination of the previously described experiments. Due to the displacement of the solid, the volume of the pores changes and an additional strain in the fluid occurs. The strain in the solid and the fluid are calculated by

$$\varepsilon^{s} = \frac{\Delta V^{s}}{V^{s}} = u_{i,i}^{s}$$

$$\varepsilon^{f} = \frac{\Delta V^{f} + \Delta V^{s}}{V^{f}} = u_{i,i}^{f} + \frac{(1-n)}{n}u_{i,i}^{s} \tag{2.8}$$

This result is in agreement with Equation (2.5).

It is noted that the kinematic interaction term is derived using the principle of conservation of mass. This could be expected, since the kinematic equations are directly related to the conservation of mass.

The grains of the solid are supposed to be infinitely stiff. The stiffness of the solid which is taken into account here, is due to changes in the position of the grains relative to the other particles.

Biot [1962, 1941] has derived a set of equations which takes into account the influence of the stiffness of the grains. Rewritten in terms of the notation adopted here these equations read

$$\frac{1}{3}\sigma^{s}_{kk} + (1-\alpha)\,\pi = K_s\, u^{s}_{k,k}$$
$$\pi = (\alpha-n)\,M\,u^{s}_{i,i} + nMu^{f}_{i,i} \tag{2.9}$$

If the grains are infinitely stiff, then $\alpha = 1$ and Equation (2.9) coincides with Equation (2.1) and (2.2), with $M = K_f/n$.

Biot has suggested to measure the elastic properties by a fully drained test and a constant volume test. In the fully drained test $\pi = 0$ and the total stress (which equals the effective stress), the solid and the fluid displacements are measured. From these results and the Equations (2.9) the solid stiffness K_s and constant α can be calculated. In the constant volume test no solid displacement may occur and the fluid stress, total stress and fluid displacement are measured. From these results and the second equation of Equation (2.9) the stiffness M can be calculated.

Equations (2.9) offer the possibility to evaluate the dependence of the parameters α and M on the stiffness of the grains K_g. Due to the compressibility of the grains an additional volumetric strain $\tilde{u}_{k,k}$ in the skeleton occurs when the fluid stress increases. In the constant volume test this volumetric strain is impossible. Therefore the skeleton will be stretched by the opposite value:

$$\tilde{u}_{k,k} = \frac{\pi}{K_g} = -\frac{\sigma^{s}_{kk}}{3K_s} \tag{2.10}$$

The constant α can be calculated by elimination of the isotropic stress by using the right-hand-side of Equation (2.10) and the first equation of Equation (2.9), taking into account that the volumetric strain of the solid is zero

$$\frac{1}{3}\sigma^{s}_{kk} = -\frac{K_s}{K_g}\pi \quad so$$
$$\frac{1}{3}\sigma^{s}_{kk} + (1-\alpha)\,\pi = -\frac{K_s}{K_g}\pi + (1-\alpha)\,\pi = 0 \tag{2.11}$$
$$\Rightarrow \alpha = 1 - \frac{K_s}{K_g}$$

For a layer close to the surface, the order of the stiffness of a sand skeleton is 20 MPa and the stiffness of a sand grain is 20 GPa, so the approximation $\alpha = 1$ is almost exact.

The stiffness M can be calculated from the constant volume test by dividing the injected fluid volume into two parts

$$u^{f}_{i,i} = \hat{u}_{i,i} + \tilde{u}_{i,i} \tag{2.12}$$

with $\hat{u}_{i,i}$ the volume needed to increase the fluid stress to the value π if the grains are incompressible

$\tilde{u}_{i,i}$ the additional volume due to the compressibility of the grains.

The volume needed to increase the fluid stress in the additional volume to the value π is neglected. These two volumes can be approximated by

$$\hat{u}_{i,i} = \pi/K_f \qquad \text{and} \qquad \tilde{u}_{i,i} = \pi/K_g$$

Substitution in the second equation of Equation (2.9) shows

$$\pi = n\,M\,u^f_{i,i} = n\,M\left(\frac{\pi}{K_f} + \frac{\pi}{K_g}\right)$$
$$\Rightarrow \qquad M = \frac{1}{n}K_f\left(\frac{1}{1 + \dfrac{K_f}{K_g}}\right) \tag{2.13}$$

The order of the stiffness of the fluid is 2 GPa and of the grains is 20 GPa, thus one finds for the order of $M = 0.9\ K_f\ /\ n$. This is a reasonable influence. However, this can be taken into account in the Equations with $\alpha = 1$ by an adjustment of the bulk modulus of the fluid.

Verruijt [1982] has derived for the same problem almost similar equations. Some differences occur. For practical values of the stiffness of the skeleton, the grains and the fluid, these differences can be neglected.

The term R_j in the constitutive equation (2.2) shows the interaction force (per volume) in the direction x_j from the fluid on the solid. This force is induced by the viscosity of the fluid when it moves through the solid. The friction is supposed to be linear, according to Darcy's Law.

The last interaction, in the dynamic equations (2.3), is described by the apparent mass, indicated by ρ_a. The apparent mass has been introduced by Biot [1956], in order to take into account the difference between the kinetic energy stored in porous medium and the kinetic energy calculated from the mean (macroscopic) fluid velocity. The difference is caused by the lateral motion of the fluid when it flows through the solid due to the irregular shape of the pores. The apparent mass increases the mass of the solid in the dynamic equation of the solid if the fluid does not move and it increases the mass of the fluid in the dynamic equation of the fluid if the solid does not move.

2.2.3 Introduction of specific discharge

The motion of the fluid can be described relative to the motion of the skeleton. Generally, the relative fluid displacement multiplied with the porosity is used in the description of dynamical behaviour of saturated porous media and the specific discharge is used in the description of static behaviour of saturated porous media. In a linear approach the (relative) fluid displacement is limited. Since during consolidation after dynamical action large fluid displacements are expected, this consolidation cannot be described with this approach. The usage of the discharge is advantageous, since large outflow from a system or large flow trough a system can be described relatively simple. It is then possible to describe the consolidation effects at larger time scales together with the dynamical action with the same algorithm.

In this section, the usage of specific discharge is adopted for the description of the dynamical behaviour.

Let u_i be the solid displacement and w_i the relative displacement

$$\begin{aligned} u_i &= u_i^s \\ w_i &= u_i^f - u_i^s \end{aligned} \tag{2.14}$$

Substitution in Equation (2.1) to (2.3) gives

$$\begin{aligned} \sigma_{ij}^s &= (K_s - \tfrac{2}{3}G_s)\, u_{k,k}\, \delta_{ij} + G_s(u_{i,j} + u_{j,i}) \\ \pi &= \frac{1}{n}K_f\, u_{k,k} + K_f\, w_{k,k} \\ \sigma_{ij,i}^s + (1-n)\pi_{,j} + b\,\dot{w}_j &= (1-n)\,\rho_s\,\ddot{u}_j - \rho_a \ddot{w}_j \\ n\pi_{,j} - b\,\dot{w}_j &= n\rho_f\,(\ddot{u}_j + \ddot{w}_j) + \rho_a\,\ddot{w}_j \end{aligned} \tag{2.15}$$

Next, the specific discharge $q_j = n\, \partial w_j / \partial t$ is introduced. The constitutive equation of the fluid is differentiated with respect to time

$$\begin{aligned} \sigma_{ij}^s &= (K_s - \tfrac{2}{3}G_s)\, u_{k,k}\, \delta_{ij} + G_s(u_{i,j} + u_{j,i}) \\ \dot{\pi} &= \frac{1}{n}K_f\,(\dot{u}_{k,k} + q_{k,k}) \\ \sigma_{ij,i}^s + (1-n)\,\pi_{,j} + \frac{b}{n}q_j &= (1-n)\rho_s\ddot{u}_j - \frac{\rho_a}{n}\dot{q}_j \\ n\pi_{,j} - \frac{b}{n}q_j &= n\rho_f\,(\ddot{u}_j + \frac{1}{n}\dot{q}_j) + \frac{\rho_a}{n}\dot{q}_j \end{aligned} \tag{2.16}$$

Furthermore, the last equation is added to the second last, showing

$$\begin{aligned} \sigma_{ij}^s &= (K_s - \tfrac{2}{3}G_s)\, u_{k,k}\delta_{ij} + G_s(u_{i,j} + u_{j,i}) \\ \dot{\pi} &= \frac{1}{n}K_f\,(\dot{u}_{k,k} + q_{k,k}) \\ \sigma_{ij,i}^s + \pi_{,j} &= [(1-n)\rho_s + n\rho_f]\,\ddot{u}_j + \rho_f\,\dot{q}_j \\ n\pi_{,j} - \frac{b}{n}q_j &= n\rho_f(\ddot{u}_j + \frac{1}{n}\dot{q}_j) + \frac{\rho_a}{n}\dot{q}_j \end{aligned} \tag{2.17}$$

In the third equation the term $(1\text{-}n)\,\rho_s + n\,\rho_f$ is the total volumetric mass of the soil and the term $\sigma^s_{ij} + \pi\,\delta_{ij}$ is the total stress. This formulation will be used for the numerical discretisation.

2.2.4 Single phase description

In order to be able to compare the description as a saturated porous material with a single phase material, the equations of motion for a single phase material are derived from the equations for the saturated porous material. Equation (2.15), (2.18) is chosen as a comfortable starting point. The relative displacement and its time derivatives are chosen zero.

$$\begin{aligned}
\sigma^s_{ij} &= (K_s - \tfrac{2}{3}G_s)\, u_{k,k}\, \delta_{ij} + G_s\,(u_{i,j} + u_{j,i}) \\
\pi &= \frac{1}{n} K_f\, u_{k,k} \\
\sigma^s_{ij,i} + (1-n)\,\pi_{,j} &= (1-n)\,\rho_s \ddot{u}_j \\
n\pi_{,j} &= n\rho_f\, \ddot{u}_j
\end{aligned} \tag{2.18}$$

Now, the second equation is added to the first one and the fourth equation is added to the third one. The total stress is introduced by $\sigma_{ij} = \sigma^S{}_{ij} + \pi\ \delta_{ij}$, and the equivalent material properties are defined as

$$\begin{aligned}
\rho_{eq} &= (1-n)\,\rho_s + n\,\rho_f \\
K_{eq} &= \frac{1}{n} K_f + K_s \\
G_{eq} &= G_s
\end{aligned} \tag{2.19}$$

This leads to the equations

$$\begin{aligned}
\sigma_{ij} &= (K_{eq} - \tfrac{2}{3}G_{eq})\, u_{k,k}\, \delta_{ij} + G_{eq}\,(u_{i,j} + u_{j,i}) \\
\sigma_{ij,i} &= \rho\ \ddot{u}_j \\
\pi &= \frac{1}{n} K_f\, u_{k,k}
\end{aligned} \tag{2.20}$$

The first two equations are just the equations of a single phase material, the third equation offers the possibility to calculate the fluid stress and next the effective stress. In literature, this material is sometimes called the frozen mixture.

2.3 Initial and boundary conditions

The equations presented in the previous sections, hold for a homogeneous field. In order to complete the description appropriate initial conditions in the field and conditions along the boundaries and along the interfaces of two different (porous) materials are required. The initial conditions and the boundary conditions are discussed in this section. The conditions at the interfaces are discussed in the next section.

2.3.1 Initial conditions

The initial conditions hold for the complete field at time $t = t_0$. First the initial conditions for the Equations (2.1), (2.2) and (2.3) are stated

$$
\begin{aligned}
u_j^s(t=t_o,x_i) &= f_j^s(x_i) \quad \textit{for the solid displacement}\\
\dot{u}_j^s(t=t_o,x_i) &= g_j^s(x_i) \quad \textit{for the solid velocity}\\
u_j^f(t=t_o,x_i) &= f_j^f(x_i) \quad \textit{for the fluid displacement}\\
\dot{u}_j^f(t=t_o,x_i) &= g_j^f(x_i) \quad \textit{for the fluid velocity}
\end{aligned}
\qquad (2.21)
$$

The functions f_j are differentiable functions.

In the previous section the field variables were changed. The initial conditions for the alternative system, Equation (2.17), are required. The specific discharge and the fluid stress are the new variables. The initial values for the alternative system can be derived from

$$
\begin{aligned}
q_j(t,x_i) &= n\dot{w}_j(t,x_i) = n\,(\dot{u}_j^f(t,x_i) - \dot{u}_j^s(t,x_i)) \Rightarrow\\
q_j(t=t_o,x_i) &= n\,(g_j^f(x_i) - g_j^s(x_i))\\
&\textit{and}\\
\pi(t,x_i) &= \frac{(1-n)}{n} K_f\, u_{k,k}^s(t,x_i) + K_f\, u_{k,k}^f(t,x_i) \Rightarrow\\
\pi(t=t_o,x_i) &= \frac{(1-n)}{n} K_f\, f_{k,k}^s(x_i) + K_f\, f_{k,k}^f(x_i)
\end{aligned}
\qquad (2.22)
$$

Equation (2.22) and the functions f_j and g_j for the solid are the complete set of initial conditions for the system of equations.

2.3.2 Boundary conditions

On the surface of the field boundary conditions are required for both the solid and the fluid. For each field variable a Dirichlet boundary condition, a von Neumann boundary condition or a mixed boundary condition is possible. For the Dirichlet boundary condition the value of the field variables is prescribed on the boundary and for the von Neumann boundary condition the derivative of the field variables is prescribed. However, these possibilities can be mixed between the field variables too. In this section some physical relevant possibilities are shown.

For the solid a prescribed displacement is called the Dirichlet boundary condition and a prescribed stress is called the von Neumann boundary condition.

For the fluid a prescribed fluid stress is called the Dirichlet Boundary condition and a prescribed specific discharge is called the von Neumann boundary condition. So, due to the change to fluid stress and specific discharge as variables, a boundary with a prescribed fluid stress changed from a von Neumann boundary to a Dirichlet boundary.

This change in terminology can be elucidated by considering Equation (2.15), (2.18) and Equation (2.17) with no solid displacement ($u_j = 0$). In the original case (Equation (2.15), (2.18)), it can be shown by substitution that if w_j and π is a set of solutions of the problem with only fluid stresses prescribed on the boundary then $w_j + A$ and π is also a set solutions of the problem. In the derived case (Equation (2.17)), it can be shown by substitution that if q_j and π is a set of solutions of the problem with only fluid stresses prescribed on the boundary then q_j and $\pi + B$ is also a set of solutions of the problem.

In dynamical studies on saturated porous media only a few boundary conditions are physical relevant:

- solid displacement and specific discharge are equal to zero
- solid stress and fluid stress are prescribed
- specific discharge is zero and a total stress is prescribed

The first case is very common, it refers to a boundary to a impermeable and infinitely stiff material. The second case refers to a soil surface which coincides with the phreatic surface. The third case relates to the loading of the soil by a (massless) impermeable membrane.

Moreover, a radiation condition might be possible. This refers to the case that waves propagate to infinity.

2.4 Conditions at interfaces

This section is dealing with the conditions of continuity at interfaces between (cf. Figure 2.1 for a sketch of the interfaces).

- two saturated porous media (interface 1)
- a saturated porous medium and an impermeable medium (interface 2)
- a saturated porous medium and a dry porous medium (interface 3)

The first and second situation have been studied by Deresiewicz and Skalak [1962]. These situations refer to a layered soil and soil-structure interaction respectively. The last situation which is discussed in this section refers to the phreatic surface.

2.4.1 Conditions between two saturated porous media

The continuity conditions at an interface between two saturated porous media will be presented. An intensive discussion on these conditions has been presented by Deresiewicz and Skalak [1963]. Here, the discussion is limited to straight interfaces.

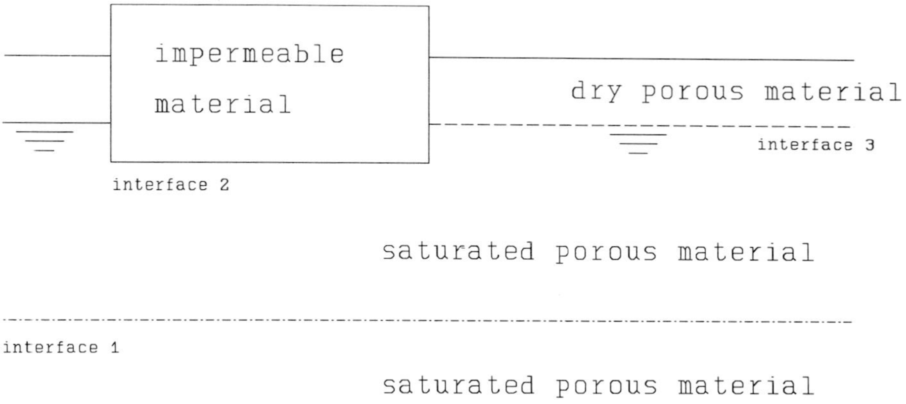

Figure 2.1 Sketch of the physical relevant interfaces

Now, both media are assumed to be saturated with the same fluid. In soil dynamics the fluid is mostly water, which might be partially filled with air. In that case the air content at both sides of the interface are assumed to be equal. Thus, the interface is completely defined by the change of properties of the solid.

Firstly the kinematic equations are studied. These are related to the principle of conservation of mass. It is assumed that both solids are tightly connected, so, no slip occurs and the displacements (and thus the velocities) of the solids at either side of the interface is continuous.

The fluid through the interface should obey the principle of conservation of mass. The fluid flow near the interface can be decomposed in the direction perpendicular to the interface and parallel to the interface. Besides the friction between the solid and the fluid, the viscosity of the fluid is not taken into account for the fluid flow. Therefore, the flow parallel with the interface might jump across the interface. For he flow perpendicular to the interface the conservation of mass shows that the specific discharge at both sides of the interface should be equal.

These considerations lead to the kinematic equations at an interface

$$
\begin{aligned}
u_n^{s_a} &= u_n^{s_b} \\
u_t^{s_a} &= u_t^{s_b} \\
q_n^{a} = q_n^{b} \;\Rightarrow\; \dot{u}_n^{f_a} - \dot{u}_n^{s_a} &= \dot{u}_n^{f_b} - \dot{u}_n^{s_a}
\end{aligned}
\qquad (2.23)
$$

The subscript n refers to the direction perpendicular to the interface, the subscript t refers to the direction parallel to the interface. Since the specific discharge equals the relative fluid velocity divided by the porosity, the fluid velocity is not continuous over the interface when the porosity of both solids is not equal.

Secondly the dynamic equations are studied. These are related to the principle of conservation of momentum. The dynamic equations cannot be derived from the dynamic equilibrium of a thin slice including the interface, because the jump in material properties may cause internal forces, which do not fade when the thickness of the slice is getting infinitely small. Deresiewicz and Skalak [1963] have employed an energy formulation in order to obtain the dynamic equations at the interface. They obtained the equations

$$\begin{aligned} \sigma_n^{s_a} + \pi^a &= \sigma_n^{s_b} + \pi^b \\ \sigma_t^{s_a} &= \sigma_t^{s_b} \\ \pi^b - \pi^a &= \frac{q_n}{\kappa_s} \end{aligned} \tag{2.24}$$

The first equation of (2.24) shows that the total stress normal at the interface must be continuous. The second equation shows that the shear stress is continuous. This is trivial when no slip occurs. The third equation shows that the pore pressure can be discontinuous, when the interface has a certain resistance against fluid flow. The physical background of the variable κ_s can be understood if one assumes that the particle size changes at the interface. The smaller particles will obstruct the pores between the bigger particles. This leads near the interface to a smaller permeability. If this decrease of permeability doesn't occur, the third equation of (2.24) shows that the pore pressure is continuous at the interface and then the first equation shows that the effective stress is continuous at the interface.

For stationary flow it can be shown that the influence of the resistance of the interface κ_s is small. According to Bourbie et al. [1987] the resistance of the interface is important for very high frequencies only.

In a stationary problem it can be derived that the specific discharge parallel with the interface obey the following relation

$$\frac{q_t^a}{K^a} = \frac{q_t^b}{K^b} \tag{2.25}$$

with q_t the parallel specific discharge
K the intrinsic permeability
The superscript refers to the material.

In a dynamical problem this is not generally true, even if the variable κ_s is infinite, see appendix 2.A.

2.4.2 Porous medium impermeable medium interface

The conditions at an interface between a saturated porous medium and an impermeable medium (soil-structure interface) will be derived from the interface conditions in the previous section, Equations (2.23) and (2.24).

Let material b be the impermeable medium. In this medium only a solid displacement and a total stress are defined. Deresiewicz and Skalak derived the following conditions

$$\sigma_n^a + \pi^a = \sigma_n^b , \quad \sigma_t^a = \sigma_t^b , \qquad \dot{u}_n^a = \dot{u}_n^b , \quad \dot{u}_t^a = \dot{u}_t^b , \quad q_n = 0 \tag{2.26}$$

2.4.3 Phreatic surface

From the general interface conditions the conditions at a phreatic surface are derived. At the phreatic surface the skeleton is continuous, so no interface permeability must be introduced: the value of $\kappa_s = \infty$. Moreover, the fluid stress in the dry part of the material, say material b, is zero. This leads to the following conditions at the interface

$$\pi^a = 0 , \quad \sigma_n^a = \sigma_n^b , \quad \sigma_t^a = \sigma_t^b , \qquad u_n^a = u_n^b , \quad u_t^a = u_t^b \tag{2.27}$$

The condition for the specific discharge in the saturated part is dropped, so it is a free variable.

In reality, the situation along a phreatic surface is more complicated. Due to capillarity, the region just above the phreatic surface is partly saturated with water. The consequences of this are quite complicated and studied by many researchers. In general, the stiffness, the permeability and the strength of the soil are influenced. In order to model these phenomena correctly, an additional state parameter, the suction, is introduced [e.g. Alonso et al., 1990]. An precise treatment of this region is outside the scope of this study.

3 ANALYTICAL SOLUTIONS OF WAVES NEAR DISCONTINUITIES

3.1 Introduction and scope of the chapter

The existence of the wave of the second kind makes the behaviour of a saturated porous medium different from a single phase material. Since the wave of the second kind is strongly damped, it is manifest mainly near discontinuities (boundaries and interfaces). Therefore, the solutions near discontinuities are studied.

This chapter is limited to the following discontinuities

- layer separation
- soil-structure
- phreatic surface.

Analytical solutions are evaluated for straight discontinuities. The problem is solved in the frequency domain. Time domain solution are found by numerical Fourier transform.

Four one-dimensional situations are considered:

- incidence of a shear wave on a layer separation
- incidence of a compressional wave of the first kind on a layer separation
- incidence of a compressional wave on a layer separation
- incidence of a compressional wave on a structure-soil interface

The analytical solution of harmonic wave propagation in saturated porous media has been solved before [e.g. Geertsma & Smit, 1961]. Their study included the interface between an impermeable solid and a saturated porous medium. However, it mainly points at the wave propagation further away from the discontinuity.

Simon et al. [1984] have created an analytical solution for the one-dimensional wave propagation in a saturated porous medium. Their solution is only valid for special materials, the so-called dynamic compatible material. Van der Grinten et al. [1987] have applied the Fourier transform to the wave propagation in saturated porous media. This solution is not restricted to the dynamic compatible material. This technic is extended to several problems with discontinuities. Recently, Degrande [1992] have applied the Fourier transform to the two-dimensional wave propagation in horizontally layered saturated porous media.

One two-dimensional problem is studied in this chapter:

- surface waves along the surface of a halfspace

The problem of a Rayleigh wave on a saturated porous medium has been initially studied by Deresiewicz [1962]. In the article, Deresiewicz paid no attention to the possibility of a second Rayleigh wave, which might exist as a composition of the shear wave and the compressional wave of the second kind. Later, Tajuddin [1984] has proposed a slightly different solution method, which offers more insight in the solution. This strategy is elaborated herein, taking into account the viscous friction between fluid and solid, which was disregarded by Tajuddin. The main target of this study is the influence of the wave of the second kind on the Rayleigh wave on a saturated porous halfspace.

Before treating the two-dimensional problem the equations from chapter 2 are rewritten in a potential form for a two-dimensional plane strain situation.

The application of Fourier transform is only possible for linear problems. In general, soils behave linearly for very low strain levels only. Therefore, the theory discussed in this chapter strictly holds for these low stress levels only. The stiffness moduli for the solid then refer to the low strain moduli, which can be measured in the field by acoustic measurement methods. However, the phenomena studied in this chapter, will also occur for non-linear soil behaviour.

3.2 One-dimensional wave propagation

3.2.1 Overview

In this section the behaviour of a saturated porous medium is studied analytical. The study is limited to one-dimensional problems. The material behaviour is linear elastic. In a one dimensional problem compressional waves and shear waves can be separated. Since the material behaves linear, the problem can be transform into the frequency domain.

Firstly, the shear wave propagation is studied. For a linear problem it is shown that the solid and the fluid move almost completely simultaneously for lower frequencies. For higher frequencies, the fluid motion depends strongly on the value of the apparent mass.

Secondly, the compressional waves are studied. Time domain solutions are calculated by applying Fourier transforms. Three cases are studied:

- wave reflection at an interface between two saturated porous media of infinite length
- wave reflection at an interface between an impermeable medium of finite length and a saturated porous medium of infinite length
- wave reflection at an interface between a saturated porous medium of finite length and a saturated porous medium of infinite length

The solution method for a step loading on a saturated porous material is firstly used by Van der Grinten et al. [1987]. The method is analytical, only the Fourier transforms are carried out numerically.

3.2.2 General solutions for propagation of shear waves in the frequency domain

Field equations

The Equations (2.1 - 2.3) in chapter 2 can be simplified for a one-dimensional problem (in x_1 = x-direction) with shear waves only

$$\begin{aligned} \tau_{,x} + b\,(v^f - v^s) &= [(1-n)\,\rho_s + \rho_a]\,\dot{v}^s - \rho_a \dot{v}^f \\ -b\,(v^f - v^s) &= -\rho_a\,\dot{v}^s + [n\,\rho_f + \rho_a]\,\dot{v}^f \\ \dot{\tau} &= G\,v^s_{,x} \end{aligned} \tag{3.1}$$

with v^s the velocity of the solid
v^f the velocity of the fluid
τ the shear stress in the solid

Substitution of harmonic waves in x-direction with an $e^{-i\kappa x}\,e^{i\omega t}$ dependence on time and x-coordinate (with the wave number κ complex ($\kappa \in \mathbb{C}$)) leads to

$$\begin{pmatrix} \omega m_2 & -\omega m_a & \kappa \\ -\omega m_a & \omega m_1 & 0 \\ \kappa G & 0 & \omega \end{pmatrix} \begin{pmatrix} \hat{v}^s \\ \hat{v}^f \\ \hat{\tau} \end{pmatrix} = \begin{pmatrix} 0 \\ 0 \\ 0 \end{pmatrix} \tag{3.2}$$

with $m_j = \rho_j - i\,b/\omega$ (j = 1, 2 and a)
$\rho_1 = n\,\rho_f + \rho_a$
$\rho_2 = (1\text{-}n)\,\rho_s + \rho_a$

For a non-trivial solution the determinant must be zero. Solving for κ leads to

$$\kappa^{\pm} = \pm\frac{\omega}{\sqrt{G}}\sqrt{\frac{m_1 m_2 - m_a^2}{m_1}} = \pm\frac{\omega}{c_s^{dry}}\sqrt{\frac{m_1 m_2 - m_a^2}{m_1\,(m_2 - m_a)}} = \pm\frac{\omega}{c_s^{dry}} f_\kappa(\omega) \tag{3.3}$$

with $c_s^{dry} = \sqrt{G/(m_2\text{-}m_a)}$ the shear wave velocity in the dry solid
the superscript + and - refer to the propagation direction.

The factor $f_\kappa(\omega)$ in Equation (3.3) shows that the relation between the shear wave velocity in dry solid and in saturated solid depends on frequency.

From the solution for the wave numbers and the original equation (3.2), the relation between the fluid velocity and the shear stress with the solid velocity are calculated

$$\frac{\hat{v}^f}{\hat{v}^s} = A^{\pm}(\omega) = \frac{m_a}{m_1}$$

$$\frac{\hat{\tau}}{\hat{v}^s} = B^{\pm}(\omega) = -\frac{\kappa^{\pm} G}{\omega} = \mp\frac{G f_k(\omega)}{c_s^{dry}} \qquad (3.4)$$

<u>Discussion on general solution</u>

Figure 3.1 shows the wave number coefficient $f_\kappa(\omega)$ and the multiplier $A(\omega)$ for a permeable sand. The following data is used: ρ_s = 2500 kg/m^3, ρ_f = 1000 kg/m^3, ρ_a = 0 kg/m^3, n = 0.4, k = 1 mm/s (or $b = n^2\rho g/k = 1.6*10^6$ N/m^3/(m/s)).

From the figure, it is concluded that for low frequencies and for high frequencies the wave number is almost real, so the shear wave propagates almost undamped. For low frequencies the wavenumber factor $f_\kappa(\omega)$ = 1.1, so the velocity of the wave is about 90 % (= $100/f_\kappa(\omega)$) of the velocity in the dry material. The solid and fluid move together since $A(\omega)$ = 1.

For intermediate frequencies the wavenumber is complex. A critical frequency exists, for which the imaginary part of the wavenumber and the multiplier $A(\omega)$ are maximum. This critical frequency can be found by the zero value of the derivative of the imaginary part of the multiplier with respect to the frequency

$$A(\omega) = \frac{m_a}{m_1} = \frac{\omega^2\rho_1\rho_a + b^2}{\omega^2\rho_1^2 + b^2} + i\,\frac{-\omega b(\rho_1 - \rho_a)}{\omega^2\rho_1^2 + b^2}$$

$$\frac{d}{d\omega}\left(\frac{\omega b(\rho_1 - \rho_a)}{\omega^2\rho_1^2 + b^2}\right) = 0 \qquad (3.5)$$

$$\Rightarrow \quad \omega = \omega_c = \frac{b}{\rho_1}$$

For this material the critical frequency is 637 Hz. The damping per meter is calculated from $1 - e^{-|Im\{\kappa\}|}$. At a frequency close to the critical frequency, a damping of approximately 5 % per meter is reached.

At frequencies far above the critical frequency, the water cannot follow the solid at all and it remains silent. The shear wave propagates undamped with the velocity of the wave in the dry solid. In this frequency region, the apparent mass is important, since

$$\lim_{\omega\to\infty} A(\omega) = \frac{\rho_a}{\rho_1} \qquad (3.6)$$

Figure 3.2 shows this in an identical picture for similar material properties, except for the apparent mass is 400 kg/m^3. The critical frequency is decreased to f_c = 318 Hz and the imaginary part of the wavenumber and multiplier A(ω) are somewhat smaller. This means that the damping decreases. In the high frequency range the multiplier A(ω) is non-zero, which means that due to apparent mass the fluid can not be silent.

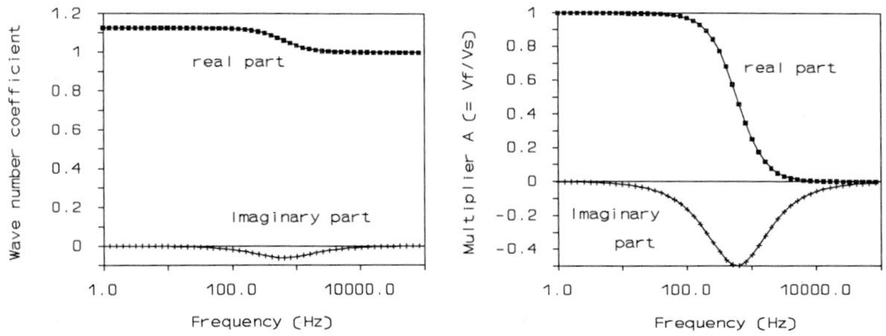

Figure 3.1 *Shear wave propagation in a permeable saturated sand (no apparent mass)*

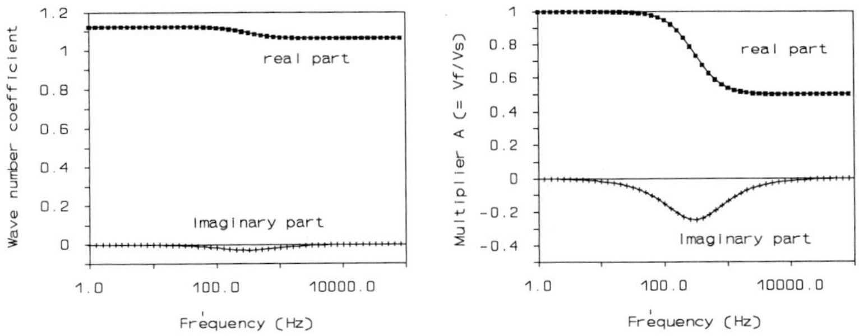

Figure 3.2 *Shear wave propagation is a permeable sand (with apparent mass)*

For practical purposes in soil mechanics the frequency can be limited to 1000 Hz. Since the critical frequency increases with the friction coefficient b, saturated sand can be considered as a single phase material for shear wave propagation in many cases.

Behaviour of shear wave at an interface

In this section the solid stresses are calculated, when a shear wave passes a straight interface between two saturated porous media. The problem is sketched in Figure 3.3. The interface is located at x_1 = 0. Material a is at negative x_1-values. A wave generated at $x_1 = -\infty$, arrives at x_1 = 0 and will be partly reflected and partly transmitted, due to different impedances of the materials a and b. Both areas are supposed to be half-infinite.

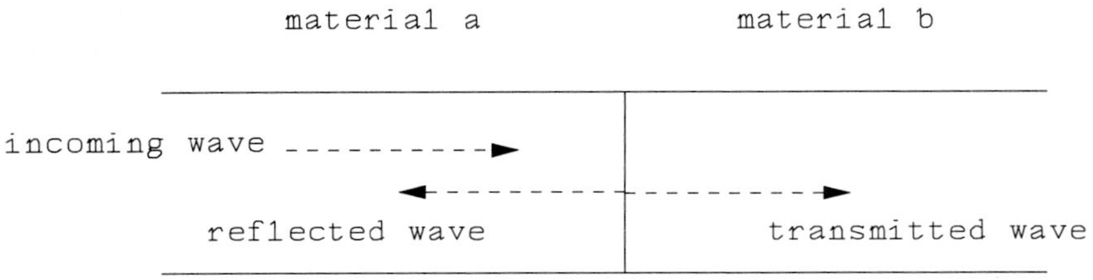

Figure 3.3 Sketch of problem for shear wave at interface

The intensity of the incoming wave is characterised by the amplitude of the solid velocity of the wave of the first kind at the interface v_0. The other variables are related to v_0 by Equation (3.4). In material 'a' two waves exist: the incoming wave and the reflected wave.

$$\begin{aligned} \hat{v}^{s,a} &= v_o + v^{s,a-} \\ \hat{v}^{f,a} &= A^{a+} v_o + A^{a-} v^{s,a-} \\ \hat{\tau}^{s,a} &= B^{a+} v_o + B^{a-} v^{s,a-} \end{aligned} \tag{3.7}$$

In material 'b' only the transmitted wave exists:

$$\begin{aligned} \hat{v}^{s,b} &= \hat{v}^{s,b+} \\ \hat{v}^{f,b} &= A^{b+} \hat{v}^{s,b+} \\ \hat{\tau}^{b} &= B^{b+} \hat{v}^{s,b+} \end{aligned} \tag{3.8}$$

Substitution in the conditions at the interface shows

$$\begin{aligned} \hat{\tau}^{a} = \hat{\tau}^{b} \quad &\Rightarrow \quad B^{a-} \hat{v}^{s,a-} + B^{a+} v_o = B^{b+} \hat{v}^{s,b+} \\ \hat{v}^{s,a} = \hat{v}^{s,b} \quad &\Rightarrow \quad \hat{v}^{s,a-} + v_o = \hat{v}^{s,b+} \end{aligned} \tag{3.9}$$

The amplitudes of the reflected and transmitted waves can be solved

$$\begin{aligned} v^{s,a-} &= \frac{B^{b+} - B^{a+}}{B^{a-} - B^{b+}} \, v_o \\ v^{s,b+} &= \frac{B^{a-} - B^{a+}}{B^{a-} - B^{b+}} \, v_o \end{aligned} \tag{3.10}$$

Since the value of the multiplier B is mainly influenced by the shear modulus of the solid and the shear wave velocity in the dry solid and hardly by the wave number coefficient $f_K(\omega)$, the shear stress close to the interface will hardly differ from the one calculated by a single phase approximation.

3.2.3 General solutions for propagation of compressional waves in the frequency domain

Field equations

The Equations in the previous chapter are simplified for a one-dimensional problem along the x_1 = x-coordinate with only compression waves. Moreover, the time derivatives of the solid displacement and the fluid displacement are replaced by the solid velocity and the fluid velocity respectively. Therefore, the constitutive equation of the fluid and solid are differentiated with respect to time.

$$\begin{aligned}
\rho_2 v^s_{,t} - \rho_a v^f_{,t} - \sigma_{,x} - (1-n)\pi_{,x} - b(v^f - v^s) &= 0 \\
-\rho_a v^s_{,t} + \rho_1 v^f_{,t} - n\pi_{,x} + b(v^f - v^s) &= 0 \\
\sigma_{,t} - K_c v^s_{,x} &= 0 \\
\pi_{,t} - K_f v^f_{,x} - \frac{1-n}{n} K_f v^s_{,x} &= 0
\end{aligned} \tag{3.11}$$

with v^s solid velocity
v^f fluid velocity
$\rho_1 = n\,\rho_f + \rho_a$
$\rho_2 = (1-n)\,\rho_s + \rho_a$
K_c constrained modulus of solid
$\sigma = \sigma^s_{11}$ normal solid stress

General solution

The field equations are solved for harmonic waves. This method has been used by other authors [e.g. Van der Kogel, 1978].
Substitution of harmonic waves $\underline{u} = \underline{\hat{u}}\, e^{-i\kappa x}\, e^{i\omega t}$ with the wave number κ complex ($\kappa \in \mathbb{C}$) leads to

$$\begin{bmatrix} \omega m_2 & -\omega m_a & \kappa & (1-n)\kappa \\ -\omega m_a & \omega m_1 & 0 & n\kappa \\ \kappa & 0 & \omega S_2 & 0 \\ \frac{1-n}{n}\kappa & \kappa & 0 & \omega S_1 \end{bmatrix} \begin{bmatrix} \hat{v}^s \\ \hat{v}^f \\ \hat{\sigma} \\ \hat{\pi} \end{bmatrix} = \begin{bmatrix} 0 \\ 0 \\ 0 \\ 0 \end{bmatrix} \tag{3.12}$$

with $m_j = \rho_j - i\, b/\omega$ (j = 1, 2 and a)
$S_1 = 1/K_f$
$S_2 = 1/K_c$

In order to find a non-trivial solution, the determinant of the matrix should be zero. This leads to a characteristic equation

$$n^2\left(\frac{\kappa^2}{\omega^2}\right)^2 - \left(n^2 m_2 S_2 + 2n(1-n)\, m_a S_2 + n m_1 S_1 + (1-n)^2 m_1 S_2 \right)\left(\frac{\kappa^2}{\omega^2}\right) + n S_1 S_2 (m_1 m_2 - m_a^2) = 0 \tag{3.13}$$

This leads to four wave numbers $\kappa_j^\pm$. The wave numbers are related to the fast and slow compressional waves (j = 1, 2) and the direction of propagation (+ or -). Once the wavenumbers κ are calculated, the eigenvectors of Equation (3.12) can be calculated. When the solid velocity is chosen as the independent variable, this results in

$$\begin{aligned} A_j^\pm &= \frac{\hat{v}_j^f}{\hat{v}_j^s} = \frac{\omega^2 m_a S_1 + (1-n)(\kappa_j^\pm)^2}{-n(\kappa_j^\pm)^2 + \omega^2 m_1 S_1} \\ B_j^\pm &= \frac{\hat{\sigma}_j}{\hat{v}_j^s} = -\frac{\kappa_j^\pm}{\omega S_2} \\ C_j^\pm &= \frac{\hat{\pi}_j}{\hat{v}_j^s} = \frac{-\omega \kappa_j^\pm [(1-n)m_1 + n m_a]}{-n^2(\kappa_j^\pm)^2 + n\omega^2 m_1 S_1} \end{aligned} \tag{3.14}$$

The subscript j refers to the kind of the wave (first or second kind) and the superscript + or - refers to the direction of propagation.

Using these results the general solution reads

$$\begin{aligned} v^s(x,t) &= (\hat{v}_1 e^{-i\kappa_1 x} + \hat{v}_2 e^{-i\kappa_2 x} + \hat{v}_3 e^{i\kappa_1 x} + \hat{v}_4 e^{i\kappa_2 x}) e^{i\omega t} \\ v^f(x,t) &= (A_1^+ \hat{v}_1 e^{-i\kappa_1 x} + A_2^+ \hat{v}_2 e^{-i\kappa_2 x} + A_1^- \hat{v}_3 e^{i\kappa_1 x} + A_2^- \hat{v}_4 e^{i\kappa_2 x}) e^{i\omega t} \\ \sigma(x,t) &= (B_1^+ \hat{v}_1 e^{-i\kappa_1 x} + B_2^+ \hat{v}_2 e^{-i\kappa_2 x} + B_1^- \hat{v}_3 e^{i\kappa_1 x} + B_2^- \hat{v}_4 e^{i\kappa_2 x}) e^{i\omega t} \\ \pi(x,t) &= (C_1^+ \hat{v}_1 e^{-i\kappa_1 x} + C_2^+ \hat{v}_2 e^{-i\kappa_2 x} + C_1^- \hat{v}_3 e^{i\kappa_1 x} + C_2^- \hat{v}_4 e^{i\kappa_2 x}) e^{i\omega t} \end{aligned} \tag{3.15}$$

This result is interpreted as follows: Each variable at position x and at time t is determined by the summation of four compressional waves. v_1 is the amplitude of the solid velocity of the fast compressional wave propagating in positive direction and v_2 is the amplitude of the solid velocity of the slow compressional wave propagating in positive direction. v_3 and v_4 are similar wave amplitudes, propagating in negative direction.

For the other variables the amplitude of each wave is calculated using the multiplier, so e.g. $B_1^+ v_1$ is the amplitude of the solid stress of the fast compressional wave propagating in positive direction.

3.2.4 Two saturated porous media of infinite length

In this section the solid stresses are calculated, when a compressional wave passes a straight interface between two saturated porous media. The problem is sketched in Figure 3.4.

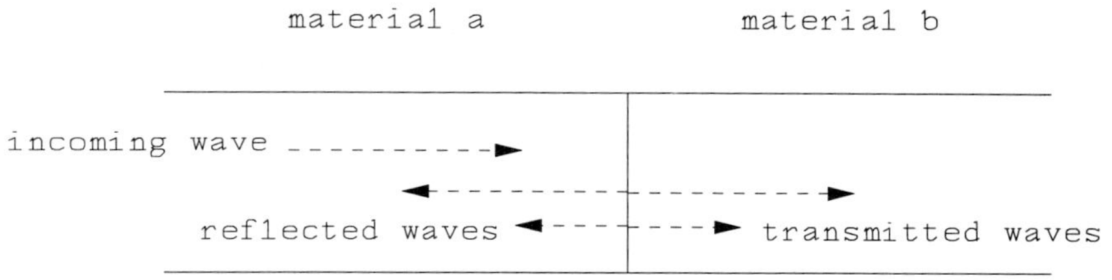

Figure 3.4 Sketch of the problem of reflection with two saturated porous materials of infinite length

The interface is located at x = 0. Material a is at negative x-values. A wave generated at x = -∞, arrives at x = 0 and will be partly reflected and partly transmitted, due to different impedances of the materials a and b. Both areas are supposed to be half-infinite.

The intensity of the incoming wave is characterised by the amplitude of the solid velocity of the wave of the first kind at the interface v_o. The other variables are related to v_o by Equation (3.15) together with Equation (3.14). Note, that only the fast wave arrives at the interface.

In material a two reflected waves are generated at the interface: the fast wave and the slow wave, propagating outward from the interface (in negative x-direction). At x = 0, the amplitudes of the field variables are defined by

$$
\begin{aligned}
\hat{v}^{s,a} &= \hat{v}_o + \hat{v}_3^a + \hat{v}_4^a \\
\hat{v}^{f,a} &= A_1^{a+}\,\hat{v}_o + A_1^{a-}\,\hat{v}_3^a + A_2^{a-}\,\hat{v}_4^a \\
\hat{\sigma}^a &= B_1^{a+}\,\hat{v}_o + B_1^{a-}\,\hat{v}_3^a + B_2^{a-}\,\hat{v}_4^a \\
\hat{\pi}^a &= C_1^{a+}\,\hat{v}_o + C_1^{a-}\,\hat{v}_3^a + C_2^{a-}\,\hat{v}_4^a
\end{aligned}
\qquad (3.16)
$$

The subscript o refers to the incoming wave, the subscripts 1 and 2 refers to the fast and the slow waves respectively. The superscripts a refers to the material and the superscript - (minus) to the propagation in negative direction and the superscript + (plus) to the propagation in positive direction.

In material b two transmitted waves are generated at the interface: the fast wave and the slow wave, propagating outward from the interface (in positive x-direction). At x = 0 the amplitudes of the field variables are defined by

$$
\begin{aligned}
\hat{v}^s &= \hat{v}_1^b + \hat{v}_2^b \\
\hat{v}^f &= A_1^{b+}\,\hat{v}_1^b + A_2^{b+}\,\hat{v}_2^b \\
\hat{\sigma} &= B_1^{b+}\,\hat{v}_1^b + B_2^{b+}\,\hat{v}_2^b \\
\hat{\pi} &= C_1^{b+}\,\hat{v}_1^b + C_2^{b+}\,\hat{v}_2^b
\end{aligned}
\tag{3.17}
$$

At the interface the following conditions hold, in accordance with the Equations (2.23) and (2.24) of chapter 2,

$$
\begin{aligned}
v^{s,b} &= v^{s,a} \\
v^{f,b} &= \frac{n^a}{n^b}\, v^{f,a} + \left(1 - \frac{n^a}{n^b}\right) v^{s,a} \\
\pi^b &= \pi^a + \frac{n^a}{\kappa_s}\left(v^{f,a} - v^{s,a}\right) \\
\sigma^b + \pi^b &= \sigma^a + \pi^a
\end{aligned}
\tag{3.18}
$$

in which κ_s is the permeability of the interface
the superscript (a or b) refers to the two materials at the interface

The field variables, defined by Equation (3.16) and (3.17), are introduced in the interface conditions (3.18). This leads to four equations with four unknowns.

$$
\begin{pmatrix}
1 & 1 & -1 & -1 \\
\eta A_1^{a-} + (1-\eta) & \eta A_2^{a-} + (1-\eta) & -A_1^{b+} & -A_1^{b+} \\
C_1^{a-} + \frac{n^a}{\kappa_s}(A_1^{a-} - 1) & C_2^{a-} + \frac{n^a}{\kappa_s}(A_2^{a-} - 1) & -C_1^{b+} & -C_2^{b+} \\
B_1^{a-} + C_1^{a-} & B_2^{a-} + C_2^{a-} & -B_1^{b+} - C_1^{b+} & -B_2^{b+} - C_2^{b+}
\end{pmatrix}
*
\begin{pmatrix}
\hat{v}_3^a \\ \hat{v}_4^a \\ \hat{v}_1^b \\ \hat{v}_2^b
\end{pmatrix}
= \hat{v}_o
\begin{pmatrix}
-1 \\
-\eta A_1^{a+} - (1-\eta) \\
-C_1^{a+} - \frac{n^a}{\kappa_s}(A_1^{a+} - 1) \\
-B_1^{a+} - C_1^{a+}
\end{pmatrix}
\tag{3.19}
$$

with $\eta = n^a/n^b$

The coefficients in the equations depend on the frequency, according to Equation (3.14). The right-hand-side of the equations is non-zero, since v_o is non-zero. From these equations all amplitudes of all waves at the interface can be solved.

Numerical example

Using the solution method presented above, it is possible to study the stress in the skeleton, when the material properties show a discontinuity. From numerical analysis, it follows that the porosity and the permeability are the most important parameters, taking notice of the practical variation of the mechanical properties of soft soils. Therefore, the materials chosen only differ in these two parameters. Material b, with low porosity and permeability, can be thought of as being more graded and denser sand than material a. Table 3.1 shows the parameters used in the numerical example.

Table 3.1 Material parameters for numerical experiment

variable		unit	mat. a	mat. b
bulkmodulus solid	K_s	MPa	12.5	12.5
shearmodulus solid	G_s	MPa	6.0	6.0
bulkmodulus fluid	K_f	MPa	2000	2000
mass solid	ρ_s	kg/m^3	2500	2500
mass fluid	ρ_f	kg/m^3	1000	1000
apparent mass	ρ_f	kg/m^3	0	0
permeability	k	mm/s	0.1	0.01
porosity	n	-	0.37	0.33
permeability interface	κ_s	(Pa s/m)$^{-1}$	∞	

The dependency of the solid stress on frequency is shown in Figure 3.6. For all frequencies the solid stress is lower than the solid stress of the incoming wave.

Figure 3.5 shows the spatial distribution around the interface for three frequencies. This shows that the reduction of stress takes place in a limited zone around the interface. This reduction is due to the existence of the wave of the second kind. The zone is limited, because the wave of the second kind is strongly damped during propagation. The damping depends on frequency, as can be seen from the decreasing size of the zone at increasing frequency.

The length of the zone can be estimated from the imaginary part from the wave number of the wave of the second kind. Van der Kogel [1977] derived analytically an approximation for the imaginary part, but they can also be calculated numerically. The wavenumber for the wave of the second kind is shown in Table 3.2, together with the with of the zone estimated by $x = 3.5/|\mathrm{Im}\{\kappa\}|$.

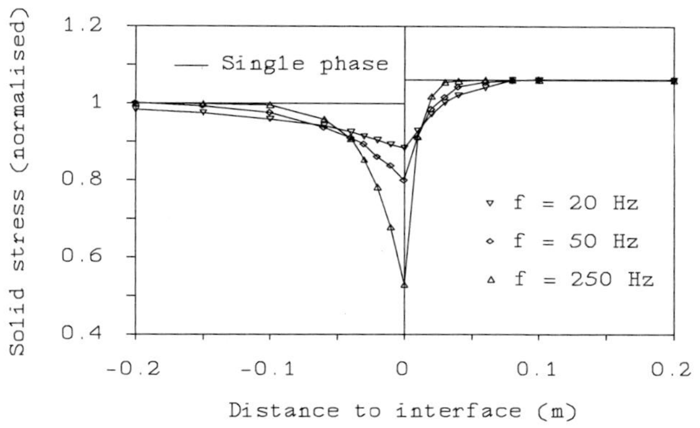

Figure 3.5 *Spatial distribution of effective stress near interface*

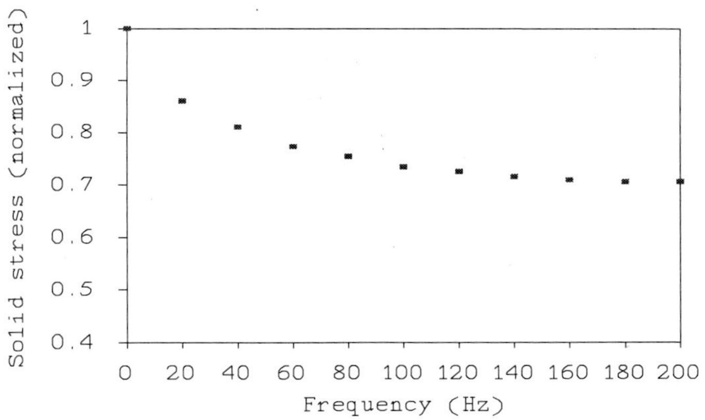

Figure 3.6 *Dependency of effective stress at interface on frequency*

3.2.5 A saturated porous material of finite length and a saturated porous material of infinite length

Now the length of material a is limited. This means that the reflections from the interface into material a will reflect at the other end of material a and reach the interface again. The problem is sketched in Figure 3.7. The interface is located at x = L. The loading exists of a step load on the pore fluid at x = 0.

A special case occurs when material a and b are equal. Then the problem of a homogeneous material is solved. This is discussed by Van der Grinten et al. [1987]. In fact, the numerical examples shown in these sections are based on the fourier transform

Table 3.2 *Imaginary part of wave numbers of wave second kind for some frequencies (material data see table above)*

	material a		material b	
frequency (Hz)	Im{κ} (m^{-1})	zone (m)	Im{κ} (m^{-1})	zone (m)
20	-18	0.19	-60	0.06
50	-28	0.13	-90	0.04
250	-60	0.06	-190	0.02

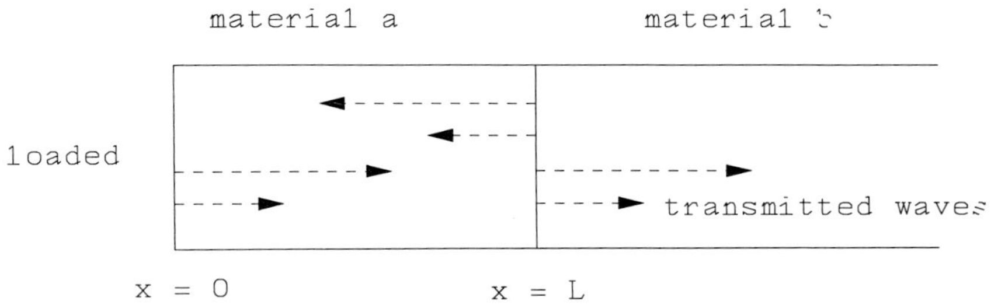

Figure 3.7 *Sketch of problem with one finite and one half infinite domain*

program written by Smeulders [1992].

The boundary and interface conditions read in this case

$$
\begin{aligned}
x = 0: \quad & \pi^a = \hat{\pi}_o \, e^{i\omega t} \\
& \sigma^a = 0 \\
x = L: \quad & v^{s,b} = v^{s,a} \\
& v^{f,b} = \frac{n^a}{n^b} v^{f,a} + \left(1 - \frac{n^a}{n^b}\right) v^{s,a} \\
& \sigma^b + \pi^b = \sigma^a + \pi^a \\
& \pi^b = \pi^a + \frac{n^a}{\kappa_s} \left(v^{f,a} - v^{s,a} \right) \\
x \to \infty \quad & \textit{radiation conditions}
\end{aligned}
\tag{3.20}
$$

For the eight unknown variables eight conditions are available. The conditions at x = 0 means that the pore water is loaded by a cyclic pressure and that the effective stress is zero. The radiation conditions can be easily fulfilled by choosing the amplitude zero for both the inward (in negative x-direction) propagating fast and slow wave.

The remaining six constants are solved by introduction of the general solutions (Equation (3.15)) (calculated for material a and material b) in the conditions (Equation (3.20)) at x = 0 and x = L. This leads to a system of six equations with six unknowns

$$\begin{pmatrix} C_1^{a+} & C_2^{a+} & C_1^{a-} & C_2^{a-} & 0 & 0 \\ B_1^{a+} & B_2^{a+} & B_1^{a-} & B_2^{a-} & 0 & 0 \\ e^{-i\kappa_1^a L} & e^{-i\kappa_2^a L} & e^{i\kappa_1^a L} & e^{i\kappa_2^a L} & -e^{-i\kappa_1^b L} & -e^{-i\kappa_2^b L} \\ F_1^{a+} e^{-i\kappa_1^a L} & F_2^{a+} e^{-\kappa_2^a L} & F_1^{a-} e^{i\kappa_1^a L} & F_2^{a-} e^{i\kappa_2^a L} & -A_1^{b+} e^{-i\kappa_2^b L} & -A_2^{b+} e^{-i\kappa_2^b L} \\ G_1^{a+} e^{-i\kappa_1^a L} & G_2^{a+} e^{-\kappa_2^a L} & G_1^{a-} e^{i\kappa_1^a L} & G_2^{a-} e^{i\kappa_2^a L} & -C_1^{b+} e^{-i\kappa_1^b L} & -C_2^{b+} e^{-i\kappa_2^b L} \\ H_1^{a+} e^{-i\kappa_1^a L} & H_2^{a+} e^{-i\kappa_2^a L} & H_1^{a-} e^{i\kappa_1^a L} & H_2^{a-} e^{i\kappa_2^a L} & -H_1^{a+} e^{-i\kappa_1^b L} & -H_2^{a+} e^{-i\kappa_2^b L} \end{pmatrix} \begin{pmatrix} \hat{v}_1^a \\ \hat{v}_2^a \\ \hat{v}_3^a \\ \hat{v}_4^a \\ \hat{v}_1^b \\ \hat{v}_2^b \end{pmatrix} = \begin{pmatrix} \hat{\pi}_o \\ 0 \\ 0 \\ 0 \\ 0 \\ 0 \end{pmatrix} \qquad (3.21)$$

$$\textit{with} \quad F_j^{ms} = \frac{n^a}{n^b} A_j^{ms} + (1 - \frac{n^a}{n^b})$$

$$G_j^{ms} = C_j^{ms} + \frac{n^m}{\kappa_s}(A_j^{ms} - 1)$$

$$H_j^{ms} = B_j^{ms} + C_j^{ms}$$

For L = 0, $v_1^a = v_o$ and $v_2^a = 0$, the last four equations of Equation (3.21) are identical with the equations in Equation (3.19).

Numerical example

The problem is solved for shock wave loading at x = 0. The rise time of the load is 1 ms. The material properties are shown in Table 3.3.

The solution for shock wave loading is calculated by using Fourier transform. The application of the Fourier transform contains three steps. The first step is the fourier transform of the loading from the time to the frequency domain. In the second step the response is calculated in the frequency domain. In this step the solution in the frequency domain (solution of Equation (3.21)), is used. The last step is the inverse Fourier transform of the response from the frequency domain to the time domain.

Figure 3.8 shows the calculated solid stresses at x = 1.00 m (the interface) and x = 1.25 m. The symbols refer to the homogeneous case, in which material a and b are identical. The lines refer to the inhomogeneous case in which the permeability is discontinue. From the homogeneous case, it is seen that the wave propagates almost undisturbed. The propagation velocity is about 1.6 m/ms.
Although the difference in permeability is only a factor 4, the solid stress for the inhomogeneous case shows great differences with the homogeneous case at the interface. However, at 1.25 m the difference is small. Obviously, this shows the influence of the compressional wave of the second kind.

Table 3.3 Material properties for examples in section 3.2.5 and 3.2.6

			section 3.2.5		section 3.2.6	
variable		unit	mat a	mat b	mat a	mat b
solid modulus	K_c	MPa	304	304	-	304
fluid modulus	K_f	MPa	2000	2000	-	2000
mass solid	ρ_s	kg/m^3	2650	2650	-	2650
mass fluid	ρ_f	kg/m^3	1000	1000	-	1000
apparent mass	ρ_a	kg/m^3	400	400	-	400
porosity	n	-	0.4	0.4	-	0.4
permeability	k	mm/s	1.0	0.25	-	1.0
permeability interface	κ_s	m/Ns	∞		-	
length	L	m	1.00		2.0	
modulus dry	K_c^d	GPa	-	-	70	-
mass dry	ρ^d	kg/m^3	-	-	7850	-

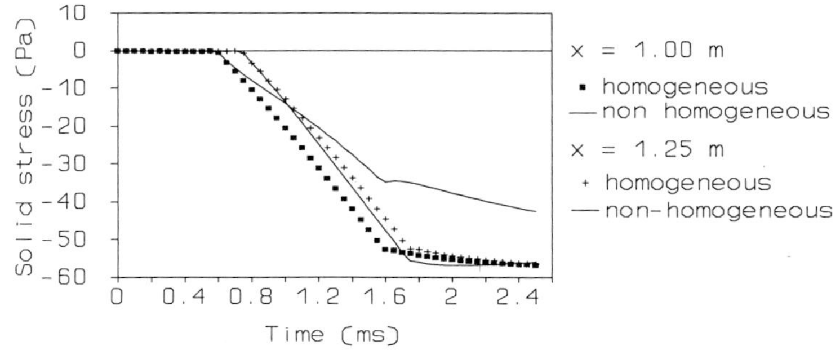

Figure 3.8 Solid stresses close to interface with jump in permeability

3.2.6 An impermeable material of finite length and a saturated porous material of infinite length

In this section the soil-structure interaction is calculated for a one-dimensional configuration. The problem is sketched in Figure 3.7. Now, material a (0 < x < L) is an impermeable single phase material. Material b (x > L) is a saturated porous material. The single phase material is loaded by a force at x = 0.

For the single phase material the solution can be derived along identical lines as mentioned before. The equations for the single phase material (Section 2.2.4) are used. The time derivative of the displacement is replaced by the velocity and the equations are simplified for the one-dimensional case

$$\begin{aligned} \sigma^d_{,t} + K^d_c v^d_{,x} &= 0 \\ \sigma^d_{,x} - \rho^d v^d_{,t} &= 0 \end{aligned} \tag{3.22}$$

An $e^{-i\kappa x}$ $e^{i\omega t}$ dependence is introduced in the equations and the wave number κ is solved. Two real wave numbers are found, one for the waves propagating in positive x-direction and one for the waves propagating in negative x-direction. This leads to the general solution in the time domain

$$\begin{aligned} v^d(x,t) &= (\hat{v}^{d+} e^{-i\kappa^d x} + \hat{v}^{d-} e^{i\kappa^d x})\, e^{i\omega t} \\ \textit{with} \quad \kappa^d &= \omega \left(\frac{K^d_c}{\rho^d} \right)^{-\frac{1}{2}} \\ &\textit{and} \\ \hat{\sigma}^{d\pm} &= \mp Z^d\, \hat{v}^{d\pm} \\ \textit{with} \quad Z^d &= \sqrt{K^d_c\, \rho^d} \end{aligned} \tag{3.23}$$

The multiplier Z^d is the impedance of the system. This variable is the multiplier for the stress in the dry material.

The boundary and interface conditions reads

$$\begin{aligned} x = 0: &\quad \sigma^d = \hat{\sigma}_o\, e^{i\omega t} \\ x = L: &\quad v^d = v^s \\ &\quad \sigma^d = \sigma^s + \pi \\ &\quad v^d = v^f \\ x \rightarrow \infty &\quad \textit{radiation conditions} \end{aligned} \tag{3.24}$$

So for the six unknowns six boundary conditions are available. The radiation conditions can be easily fulfilled by choosing the amplitude zero for both the fast wave and the slow wave. The remaining four constants are solved from

$$\begin{pmatrix} Z^d & -Z^d & 0 & 0 \\ e^{\kappa^d L} & e^{-\kappa^d L} & -e^{\kappa_1 L} & -e^{\kappa_2 L} \\ 0 & 0 & A_1^+ e^{\kappa_1 L} & A_2^+ e^{\kappa_2 L} \\ Z^d e^{\kappa^d L} & -Z^d e^{-\kappa^d L} & -(B_1^+ + C_1^+) e^{\kappa_1 L} & -(B_2^+ + C_2^+) e^{\kappa_2 L} \end{pmatrix} \begin{pmatrix} \hat{v}^{d+} \\ \hat{v}^{d-} \\ \hat{v}_1^s \\ \hat{v}_2^s \end{pmatrix} = \begin{pmatrix} \hat{\sigma}_o \\ 0 \\ 0 \\ 0 \end{pmatrix} \tag{3.25}$$

Numerical example

As an example a problem is solved for shock wave loading at x = 0. The rise time of the load is 1 ms, thereafter the load is constant. The material properties are shown in Table 3.3. The length of the impermeable material is 2.0 m. The response on the shock wave are calculated with Fourier transforms.

Figure 3.9 shows the solid stresses at x = 2.0 m (at the interface) and x = 2.2 m (20 cm from the interface). The full lines are calculated by using the Biot model, the dotted lines are calculated by using the single phase approximation. The lower part of the Figure is a blow-up of the first 6 ms of the curves in the upper part.

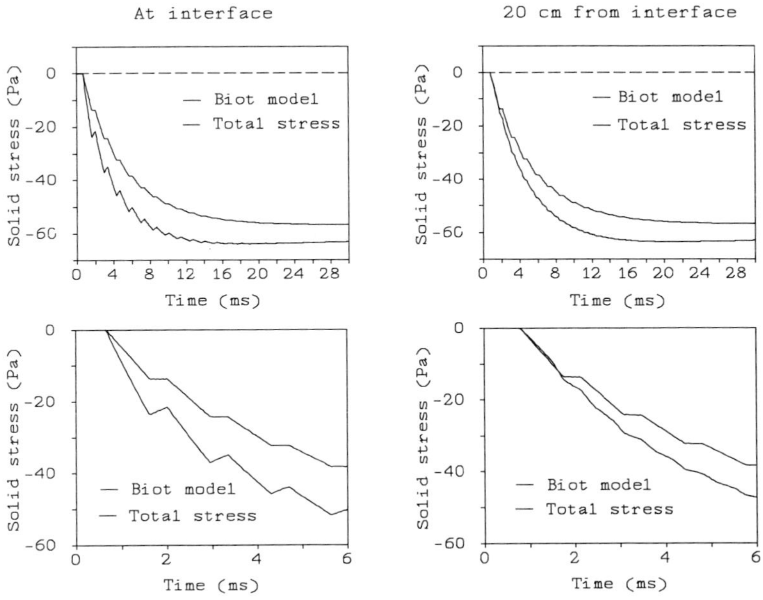

Figure 3.9 *Solid stress close to an interface with an impermeable material*

Two differences between the solution by using the Biot theory and the equivalent single phase material can be seen

- the Biot theory gives higher solid stresses (up to 12%)
- the Biot theory gives a ratchetted curve at the interface, but at 20 cm from the interface the curve is much smoother.

In order to elucidate these differences, the solid stress is examined more carefully in Figure 3.10. Figure 3.10a shows the solid stress at the interface, at 10 cm and 20 cm from the interface. The curves at the interface and at 20 cm from the interface are identical to the curves in Figure 3.9.

Point A0 on curve 0 shows the arrival of the stress wave at the interface ($t = \tau = L/c^{dry} = 2/2.986 = 0.67$ ms). Point C0 is 1 ms after A0, this is the end of the rising of the loading. The wave is partly reflected at the interface. After reflection at the loaded end at $x = 0.0$ m, the so-called first reflection arrives at the interface. Point E0 shows the arrival of this first reflection ($t = 3\tau = 0.201$ ms). Point G0 shows the end of rising of the first reflection, 1 ms after E0.

The points A1 and A2 show the arrivals of the fast wave at two positions in the saturated porous material at 0.1 m and 0.2 m from the interface, respectively. The wave propagation velocity is approximately 1.6 m/ms. The points C1 and C2 are 1 ms after A1 and A2 respectively, showing the arrival of the end of rising of the loading.

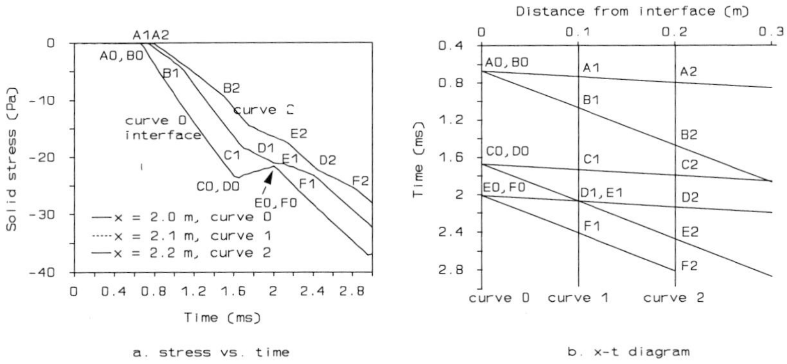

Figure 3.10 *a. Solid stress at the interface and at close to the interface*
b. x-t diagram for waves close to the interface

The kinks at the points B1 and B2 don't occur in the single phase approximation nor at the interface. After these kinks, the steepness of the curves (1 and 2) is equal to the steepness of the curve at the interface (curve 0). The time, elapsed after A0 is proportional to the distance to the interface.

Obviously, a second wave is generated at the interface. Based on the steepness of the curves before and after the kinks, the magnitude of the stress in the first wave is approximately 53% of the stress at the interface, and the magnitude of the stress in the second wave is approximately 47% of the stress at the interface.

From these observations, an x-t diagram is constructed in Figure 3.10b. The characteristic of the initial wave (full lines, A and B), the end of rising (dashed lines, C and D) and the first reflection (dotted lines, E and F) are shown for both waves (flat and steep lines respectively). This x-t diagram is used to explain the curves in Figure 3.10a. The points C1 and C2 are indeed related to the first wave. However, the solid stress doesn't decrease (as it does at the interface), since the magnitude of the first wave is not big enough. At x = 2.1 m the arrival of the second wave end of rising (point D1) and the arrival of the first wave of the first reflection (point E1) cannot be distinguished in Figure 3.10b, since these points almost coincides in time. At x = 0.2 m, the first wave of the first reflection (point E2) has overtaken the second wave of the end of rising (point D2), so these points can be seen (however, they are changed in order).

It is concluded that a wave of the second kind is generated at an interface between an impermeable solid and a saturated porous material. The occurrence of the wave of the second kind is different from the findings of Geertsma and Smit [1961].

3.3 Rayleigh wave on saturated porous halfspace

3.3.1 Overview

This Section answers the question whether one or two Rayleigh waves exists on a saturated porous halfspace. The question can be raised, since a Rayleigh wave is a combination of a shear wave and a compression wave. In a saturated porous medium two compressional waves exist, so two Rayleigh waves might be expected. This has not been discussed in literature.

First the Equations derived in Chapter 2 are rewritten in potential form for plane strain, using a Helmholtz decomposition. Thereafter, the general solution for plane wave propagation is derived. The surface waves can be found by considering the boundary conditions at the surface. This leads to the solution.

3.3.2 Potential description

The potential description will be derived from the equations (2.1), (2.2) and (2.3) in Chapter 2. Choosing the displacements as fundamental unknown one can rewrite the equations to

$$\begin{aligned} &[(1-n)\,\rho_s + \rho_a]\,\ddot{u}_j^s - \rho_a\,\ddot{u}_j^f = \\ &(K_s - \tfrac{2}{3}G_s)\,u_{k,kj}^s + G_s\,(u_{i,ij}^s + u_{j,ii}^s) + \\ &(1-n)\,K_f\,(u_{i,ij}^f + \tfrac{1-n}{n}\,u_{i,ij}^s) + b(\dot{u}_j^f - \dot{u}_j^s) \end{aligned} \qquad (3.26)$$

$$\begin{aligned} &-\rho_a\,\ddot{u}_j^s + [n\rho_f + \rho_a]\,\ddot{u}_j^f = \\ &n\,K_f\,u_{i,ij}^f + (1-n)\,K_f\,u_{i,ij}^s - b(\dot{u}_j^f - \dot{u}_j^s) \end{aligned} \qquad (3.27)$$

Secondly, the Helmholtz decomposition is made. For a general three dimensional problem, a vector $\underline{a}$ is decomposed in two independent potentials: a scalar potential φ and a vector potential $\underline{\psi}$

$$\underline{a} = \underline{\nabla}\phi + \underline{\nabla}*\underline{\psi}$$

$$\Leftrightarrow$$

$$\begin{pmatrix} a_1 \\ a_2 \\ a_3 \end{pmatrix} = \begin{pmatrix} \frac{\partial\phi}{\partial x_1} \\ \frac{\partial\phi}{\partial x_2} \\ \frac{\partial\phi}{\partial x_3} \end{pmatrix} + \begin{pmatrix} \frac{\partial\psi_2}{\partial x_3} - \frac{\partial\psi_3}{\partial x_2} \\ \frac{\partial\psi_1}{\partial x_3} - \frac{\partial\psi_3}{\partial x_1} \\ \frac{\partial\psi_1}{\partial x_2} - \frac{\partial\psi_2}{\partial x_1} \end{pmatrix} \tag{3.28}$$

with $\underline{\nabla}$ the nabla operator
and $*$ the vector product

When the problem is two-dimensional, the third component of $\underline{a}$ is zero and the derivatives in the x_3-direction are zero. Then the decomposition is defined by

$$\begin{aligned} a_1 &= \frac{\partial\phi}{\partial x_1} + \frac{\partial\psi_3}{\partial x_2} \\ a_2 &= \frac{\partial\phi}{\partial x_2} - \frac{\partial\psi_3}{\partial x_1} \end{aligned} \tag{3.29}$$

The potentials ψ_1 and ψ_2 can be chosen zero without loss of generality.

From now on the notation will be simplified. The component ψ_3 of the vector potential will be called ψ and the derivatives of the potential to direction x_i will be written by a subscript i.

For both the fluid and the solid two potentials are introduced, accordingly to Equation (3.29)

$$\begin{aligned} u_1^s &= \phi_{,1}^s + \psi_{,2}^s \\ u_2^s &= \phi_{,2}^s - \psi_{,1}^s \\ u_1^f &= \phi_{,1}^f + \psi_{,2}^f \\ u_2^f &= \phi_{,2}^f - \psi_{,1}^f \end{aligned} \tag{3.30}$$

These equations can be introduced in Equations (3.26) and (3.27) with j = 1 and 2.

By differentiation of the equations in x_1-direction with respect to x_1 and the equation in x_2-direction with respect to x_2 and summation of the results, the potential equations for the scalar potentials φ^s and φ^f are obtained

$$[(1-n)\,\rho_s + \rho_a]\,\ddot{\phi}^s - \rho_a\,\ddot{\phi}^f - (K_s + \tfrac{4}{3}G_s)\,\Delta\phi^s - \frac{(1-n)^2}{n}K_f\,\Delta\phi^s - (1-n)\,K_f\,\Delta\phi^f - b(\dot{\phi}^f + \dot{\phi}^s) = 0 \tag{3.31}$$

$$(n\,\rho_f + \rho_a)\,\ddot{\phi}^f - \rho_a\,\ddot{\phi}^s - n\,K_f\,\Delta\phi^f - (1-n)\,K_f\,\Delta\phi^s - b(\dot{\phi}^f - \dot{\phi}^s) = 0 \tag{3.32}$$

with Δ the Laplace operator.

In this derivation, the fact that $\Delta f = 0$ is sufficient condition for $\Delta\Delta f = 0$ is used.

By differentiation of the equations in x_1-direction with respect to x_2 and the equation in x_2-direction with respect to x_1 and subtraction of the results, the potential equations for the vector potentials ψ^s and ψ^f are obtained

$$[(1-n)\,\rho_s + \rho_a]\,\ddot{\psi}^s - \rho_a\,\ddot{\psi}^f - G_s\,\Delta\psi^s - b(\dot{\psi}^f - \dot{\psi}^s) = 0 \tag{3.33}$$

$$(n\,\rho_f + \rho_a)\,\ddot{\psi}^f - \rho_a\,\ddot{\psi}^s + b(\dot{\psi}^f - \dot{\psi}^s) = 0 \tag{3.34}$$

In order to simplify the notation some new variables are introduced

$$\begin{aligned} P &= K_s + \tfrac{4}{3}G_s + \frac{(1-n)^2}{n}K_f \\ Q &= (1-n)\,K_f \\ R &= n\,K_f \\ N &= G_s \\ \rho_1 &= (1-n)\,\rho_s - \rho_a \\ \rho_2 &= n\,\rho_f - \rho_a \end{aligned} \tag{3.35}$$

leading to the following potential equations

$$\begin{aligned} P\Delta\phi^s + Q\Delta\phi^f &= \rho_1\ddot{\phi}^s - \rho_a\ddot{\phi}^f - b(\dot{\phi}^f - \dot{\phi}^s) \\ Q\Delta\phi^s + R\Delta\phi^f &= -\rho_a\ddot{\phi}^s + \rho_2\ddot{\phi}^f + b(\dot{\phi}^f - \dot{\phi}^s) \\ N\Delta\psi^s &= \rho_1\ddot{\psi}^s - \rho_a\ddot{\psi}^f - b(\dot{\psi}^f - \dot{\psi}^s) \\ 0 &= -\rho_a\ddot{\psi}^s + \rho_2\ddot{\psi}^f + b(\dot{\psi}^f - \dot{\psi}^s) \end{aligned} \tag{3.36}$$

This is a set of four partial differential equations for the four field variables φ^s, φ^f, ψ^s and ψ^f. They can be solved by using harmonic analysis.

3.3.3 General solution for plane waves

The analysis used in this section is based on the approach of Tajuddin [1984]. Plane harmonic waves which propagate in the x_2-direction are assumed

$$\begin{aligned} \phi^s &= f_s(x_1)\ \exp[\,i\,(\kappa x_2 + \omega\, t\,)] \\ \phi^f &= f_f(x_1)\ \exp[\,i\,(\kappa x_2 + \omega\, t\,)] \\ \psi^s &= g_s(x_1)\ \exp[\,i\,(\kappa x_2 + \omega\, t\,)] \\ \psi^f &= g_f(x_1)\ \exp[\,i\,(\kappa x_2 + \omega\, t\,)] \end{aligned} \tag{3.37}$$

with κ wave number
and ω angular frequency

Substitution in the potential equations (3.36), lead to two systems of equations

$$\begin{aligned} &P\,(\,f_s^{II} - \kappa^2 f_s\,) + Q\,(\,f_f^{II} - \kappa^2 f_f\,) = \\ &\qquad -\omega^2\,(\,m_{11} f_s + m_{12} f_f\,) \\ &Q\,(\,f_s^{II} - \kappa^2 f_s\,) + R\,(\,f_f^{II} - \kappa^2 f_f\,) = \\ &\qquad -\omega^2\,(\,m_{12} f_s + m_{22} f_f\,) \end{aligned} \tag{3.38}$$

$$\begin{aligned} N\,(\,g_s^{II} - k^2\,g_s\,) &= -\omega^2\,(\,m_{11}\,g_s + m_{12}\,g_f\,) \\ 0 &= -\omega^2\,(\,m_{12}\,g_s + m_{22}\,g_f\,) \end{aligned} \tag{3.39}$$

with

$$\begin{aligned} m_{11} &= \rho_1 - i\frac{b}{\omega} \\ m_{22} &= \rho_2 - i\frac{b}{\omega} \\ m_{12} &= -\rho_a + i\frac{b}{\omega} \end{aligned} \tag{3.40}$$

The derivatives with respect to x_2 are written by using superscript roman numbers.

<u>Scalar potential</u>

Firstly, the equations for the scalar potentials φ^s and φ^f, Equation (3.38), are considered. The second derivative of f_f is eliminated by multiplying the first Equation by R and the second by Q. After substraction one finds

$$\begin{aligned} &(PR - Q^2)\ (f_s^{II} - \kappa^2 f_s) = \\ -\omega^2\,(Rm_{11} - Qm_{12})\,f_s &- \omega^2\,(Rm_{12} - Qm_{22})\,f_f \end{aligned} \tag{3.41}$$

and the function f_f can be expressed in f_s and its derivative

$$f_f = H_1^2\,f_s^{II} - H_2^2\,f_s \tag{3.42}$$

with

$$H_1^2 = \frac{-(PR - Q^2)}{\omega^2\,(Rm_{12} - Qm_{22})}$$
$$H_2^2 = \frac{Rm_{11} - Qm_{12}}{Rm_{12} - Qm_{22}} + \kappa^2 H_1^2 \tag{3.43}$$

This equation can be used for the elimination of the second derivative of f_f from the first equation of Equation (3.38). This leads to a fourth order equation

$$QH_1^2 f_s^{IV} + [P - QH_2^2 + \omega^2 m_{12} H_1^2 - \kappa^2 QH_1^2] f_s^{II}$$
$$+ [\kappa^2 QH_2^2 - \kappa^2 P + \omega^2 m_{22} - \omega^2 m_{12} H_2^2] f_s = 0 \tag{3.44}$$

Now, the general solution can be found by writing this fourth order equation in a simplified form

$$(\frac{d^2}{dx_1^2} + \omega^2 E_1^2)\,(\frac{d^2}{dx_1^2} + \omega^2 E_2^2)\, f_s = 0$$
$$\Rightarrow$$
$$f_s^{IV} + \omega^2 (E_1^2 + E_2^2)\, f_s^{II} + \omega^4 E_1^2 E_2^2\, f_s = 0 \tag{3.45}$$

This is compatible with Equation (3.44) if the following holds

$$QH_1^2\, \omega^2 (E_1^2 + E_2^2) = P - QH_2^2 + \omega^2 m_{12} H_1^2 - \kappa^2 QH_1^2$$
$$QH_1^2\, \omega^4 E_1^2 E_2^2 = \kappa^2 H_2^2 Q - \kappa^2 P + \omega^2 m_{22} - \omega^2 m_{12} H_2^2 \tag{3.46}$$

Now, it is recalled that the solutions of a quadratic equation a x^2 + b x + c = 0 are related by $x_1 + x_2 = -b/a$ and $x_1 x_2 = c/a$. After multiplication of the first equation of (3.46) with ω^2, it can be seen that E_3^2 and E_4^2 are the solution of a quadratic equation

$$a\,x^2 + b\,x + c = 0 \tag{3.47}$$

with

$$a = \omega^4\, Q\, H_1^2$$
$$b = -\omega^2 (P - Q H_2^2 + \omega^2 m_{12} H_1^2 - \kappa^2 Q H_1^2)$$
$$c = \kappa^2 Q H_2^2 - \kappa^2 P + \omega^2 m_{11} - \omega^2 m_{12} H_2^2 \tag{3.48}$$

The function f_s can be solved from Equation (3.45)

$$f_s^{II} + \omega^2 E_1^2 f_s = 0 \quad \underline{or} \quad f_s^{II} + \omega^2 E_2^2 f_s = 0 \tag{3.49}$$

The general solution of these equation is

$$f_s = A_1 e^{i\omega E_1 x_1} + A_2 e^{i\omega E_2 x_1} + A_3 e^{-i\omega E_1 x_1} + A_4 e^{-i\omega E_2 x_1} \tag{3.50}$$

This solution shows that there are two waves in positive direction and two waves in negative direction.

We are interested in waves, which decays in negative x_1 direction. This is only possible if E_1 and E_2 have a positive imaginary part and A_3 and A_4 are zero. Thus the solution for the function f_s reads

$$\begin{aligned} f_s &= A_1\, e^{i\omega E_1 x_1} + A_2\, e^{i\omega E_2 x_1} \\ \textit{with} \quad & Im(E_1) > 0 \;\; \textit{and} \;\; Im(E_2) > 0 \end{aligned} \tag{3.51}$$

The function f_f is related to f_s by Equation (3.42). Substitution of f_s shows

$$\begin{aligned} f_f &= B_1 A_1\, e^{i\omega E_1 x_1} + B_2 A_2\, e^{i\omega E_2 x_1} \\ & \textit{with} \\ B_1 &= -\omega^2 E_1^2 H_1^2 - H_2^2 \\ B_2 &= -\omega^2 E_2^2 H_1^2 - H_2^2 \end{aligned} \tag{3.52}$$

Thus the solution for the potential reads

$$\begin{aligned} \phi_s &= f_s\, e^{i(\kappa x_2 + \omega t)} = \\ & A_1\, e^{i\omega E_1 x_1}\, e^{i(\kappa x_2 + \omega t)} + A_2\, e^{i\omega E_2 x_1}\, e^{i(\kappa x_2 + \omega t)} \end{aligned} \tag{3.53}$$

$$\begin{aligned} \phi_f &= f_f\, e^{i(\kappa x_2 + \omega t)} = \\ & B_1 A_1\, e^{i\omega E_1 x_1}\, e^{i(\kappa x_2 + \omega t)} + B_2 A_2\, e^{i\omega E_2 x_1}\, e^{i(\kappa x_2 + \omega t)} \end{aligned} \tag{3.54}$$

<u>Vectorial potential</u>

The vectorial potentials ψ^s and ψ^f are derived by using a similar procedure, however, this is simpler. From the second equation of (3.39) one finds

$$g_f = -\frac{m_{12}}{m_{22}}\, g_s \tag{3.55}$$

Substitution in the first equation gives

$$\begin{aligned} g_s^{II} + \omega^2 E_3^2\, g_s &= 0 \\ & \textit{with} \\ E_3^2 &= -\left(\frac{\kappa^2}{\omega^2} - \frac{1}{N}\, \frac{m_{11} m_{22} - m_{12}^2}{m_{22}} \right) \end{aligned} \tag{3.56}$$

Taking into account only those waves which decay in negative x_1 direction (thus $Im(E_s) > 0$), one finds

$$\begin{aligned} \psi_s &= A_3\, e^{i\omega E_3 x_1}\, e^{i(\kappa x_2 + \omega t)} \\ \psi_f &= B_3\, \psi_s \quad \textit{with} \quad B_3 = -\frac{m_{12}}{m_{22}} \end{aligned} \tag{3.57}$$

This result is in accordance with the results in Section 3.2.2.

3.3.4 Boundary conditions

At the free surface of the saturated porous halfspace, three conditions hold: no shear stress, no normal stress and no fluid stress. These conditions must be reformulated in potential description. For computational convenience, the conditions for normal stress and fluid stress are rewritten.

By using the Equations (2.1) and (2.2) of chapter 2 and the Equations (3.30) and (3.35) of this chapter, the shear stress can be expressed by

$$\begin{aligned} \tau &= G_s\,(u^s_{1,2} + u^s_{2,1}) = \\ &N\,[\,2\,\phi^s_{,12} - \psi^s_{,11} + \psi^s_{,22}\,] = 0 \end{aligned} \tag{3.58}$$

The fluid stress is multiplied with the porosity and by using the Equations (2.1) and (2.2) of chapter 2 and the Equations (3.30) and (3.35) of this chapter, written as

$$\begin{aligned} n\,\pi &= n\,K_f(u^f_{1,1} + u^f_{2,2}) + (1-n)\,K_f(u^s_{1,1} + u^s_{2,2}) = \\ &R\,(\phi^f_{,11} + \phi^f_{,22}) + Q\,(\phi^s_{,11} + \phi^s_{,22}) = 0 \end{aligned} \tag{3.59}$$

The normal stress is rewritten by using the Equations (2.1) and (2.2) of chapter 2 and the Equations (3.30) and (3.35) of this chapter

$$\begin{aligned} \sigma_{11} &= (K_s - \tfrac{2}{3}G_s)\,(u^s_{1,1} + u^s_{2,2}) + 2\,G_s\,u^s_{1,1} = \\ &(K_s + \tfrac{4}{3}\,G_s + \tfrac{(1-n)^2}{n}\,K_f)\,(u^s_{1,1} + u^s_{2,2}) \\ &- \tfrac{(1-n)^2}{n}\,K_f(u^f_{1,1} + u^f_{2,2}) - 2G_s\,u^s_{2,2} = 0 \end{aligned} \tag{3.60}$$

By using Equation (3.59) the second term can be replaced by (1-n) K_f $(u^f_{1,1}+u^f_{2,2})$, leading to

$$\begin{aligned} \sigma_{11} &= P\,(\phi^s_{,11} + \phi^s_{,22}) + \\ &Q\,(\phi^f_{,11} + \phi^f_{,22}) - 2N\,(\phi^s_{,11} + \psi^s_{,21}) = 0 \end{aligned} \tag{3.61}$$

3.3.5 Solution

The general solutions, Equation (3.53), (3.54) and (3.57), contain three unknown constants, A_1, A_2 and A_3. These constants can be calculated from the requirement that the solution satisfies the boundary conditions at the surface. Therefore, the general solutions are introduced in the boundary conditions with x_1 = 0. This leads to three equations with three unknowns

$$\begin{pmatrix} C_{11} & C_{12} & C_{13} \\ C_{21} & C_{22} & C_{23} \\ C_{31} & C_{32} & C_{33} \end{pmatrix} \begin{pmatrix} A_1 \\ A_2 \\ A_3 \end{pmatrix} = \begin{pmatrix} 0 \\ 0 \\ 0 \end{pmatrix} \tag{3.62}$$

with

$$\begin{aligned} C_{11} &= 2N\kappa^2 - (\kappa^2 + \omega^2 E_1^2)(P + QB_1) \\ C_{21} &= N\kappa\,\omega\,E_1 \\ C_{31} &= (\kappa^2 + \omega^2 E_1^2)(Q + RB_1) \\ C_{12} &= 2N\kappa^2 - (\kappa^2 + \omega^2 E_2^2)(P + QB_2) \\ C_{22} &= N\kappa\,\omega\,E_2 \\ C_{32} &= (\kappa^2 + \omega^2 E_2^2)(Q + RB_2) \\ C_{13} &= -\omega\,\kappa N E_3 \\ C_{23} &= (\kappa^2 + \omega^2 E_3^2) N \\ C_{33} &= 0 \end{aligned} \tag{3.63}$$

The system in Equation (3.62) is homogeneous. In order to find a non-trivial solution, the determinant of the matrix in Equation (3.62) must be zero. So the wave number κ which gives a zero determinant must be sought. Once the wave number is known, the relation between the amplitudes A_1, A_2 and A_3 can be calculated.

The velocity and the damping of the waves can be calculated from the real and imaginary part of the wave number κ.

Finally, the solid and fluid displacements can be calculated, by substitution of the solution in Equation (3.30). The displacements are

$$\begin{aligned} u_1^s(x_1,x_2) &= [i\omega E_1 A_1 e^{i\omega E_1 x_1} + i\omega E_2 A_2 e^{i\omega E_2 x_1} + \\ &\quad i\kappa A_3 e^{i\omega E_3 x_1}]\, e^{i(\kappa x_2 + \omega t)} \\ u_2^s(x_1,x_2) &= [i\kappa A_1 e^{i\omega E_1 x_1} + i\kappa A_2 e^{i\omega E_2 x_1} - \\ &\quad i\omega E_3 A_3 e^{i\omega E_3 x_1}]\, e^{i(\kappa x_2 + \omega t)} \end{aligned} \tag{3.64}$$

$$\begin{aligned} u_1^f(x_1,x_2) &= [i\omega E_1 B_1 A_1 e^{i\omega E_1 x_1} + i\omega E_2 B_2 A_2 e^{i\omega E_2 x_1} + \\ &\quad i\kappa B_3 A_3 e^{i\omega E_3 x_1}]\, e^{i(\kappa x_2 + \omega t)} \\ u_2^f(x_1,x_2) &= [i\kappa B_1 A_1 e^{i\omega E_1 x_1} + i\kappa B_2 A_2 e^{i\omega E_2 x_1} - \\ &\quad i\omega E_s B_s A_3 e^{i\omega E_3 x_1}]\, e^{i(\kappa x_2 + \omega t)} \end{aligned} \tag{3.65}$$

3.3.6 Numerical example

<u>Number of surface waves</u>

A zero determinant of Equation (3.62) is found for several wave numbers κ. However, only one of these wave numbers can be identified with a real surface wave. A real surface wave propagates along the surface, i.e. in x_2-direction. From the Equations (3.53), (3.54) and (3.57), it is seen that this is only possible if the real part of E_1, E_2 and E_3 is zero or if the corresponding amplitude A_1, A_2 or A_3 is zero.

Table 3.4 Material properties for surface wave

property		value	unit
bulk modulus solid	K_s	50	MPa
shear modulus solid	G_s	6	MPa
bulk modulus fluid	K_f	2000	MPa
volumetric mass solid	ρ_s	2500	kg/m^3
volumetric mass fluid	ρ_f	1000	kg/m^3
apparent volumetric mass	ρ_a	0	kg/m^3
porosity	n	0.3	-
friction coefficient (permeability	b k	$14.4*10^6$ $9*10^{-4}$	$N/(m/s)/m^3$ m/s)

The surface wave is calculated for a specific material. The material properties are presented in Table 3.4. The frequency is 10 Hz.

Table 3.5 shows the result. Since the solution of Equation (3.62) leads to a dependent system, only the relation between A_1, A_2 and A_3 can be calculated. Therefore, the A-values in Table 3.5 are relative values. From this table it is concluded that the amplitude A_1 is much smaller than the amplitudes A_2 and A_3 and that E_2 and E_3 are indeed almost imaginary numbers. This solution represents a surface wave. The celerity is $c_R = 2\pi f/\mathrm{Re}(\kappa) = 52.4$ m/s. Since this is the only solution of Equation (3.62) which represents a surface wave, it is concluded that only one single surface wave occurs on a saturated porous halfspace, irrespective of the existence of two compressional waves.

Table 3.5 Solution for surface wave (f = 10 Hz)

variable	symbol	real part	imaginary part
wave number	κ	1.200	$-0.5455*10^{-3}$
amplitude	A_1	$-0.4232*10^{-3}$	$0.0026*10^{-3}$
	A_2	$2.948*10^{-3}$	$0.5349*10^{0}$
	A_3	$1.134*10^{0}$	$0.0000*10^{0}$
E-values	E_1	$-0.1483*10^{0}$	$0.1493*10^{0}$
	E_2	$0.0086*10^{-3}$	$0.01909*10^{0}$
	E_3	$0.0278*10^{-3}$	$4.786*10^{-3}$
B-values	B_1	-2.362	$0.0118*10^{-3}$
	B_2	1.000	$1.351*10^{-3}$
	B_3	1.000	$-1.309*10^{-3}$

Comparison with literature

Bourbie et al. [1985] presented a solution for the surface wave velocity and the damping. Their solution is for the velocity

$$2\sqrt{2}\,\frac{m^2 n}{a}\,\frac{c_R}{c_p}\,[1-(\frac{c_r}{c_s})^2]^{\frac{1}{2}}\,(\frac{\omega}{\omega_c})^{\frac{1}{2}} = 4\,[1-(\frac{c_R}{c_s})^2]^{\frac{1}{2}}\,[1-(1\frac{c_r}{c_p})^2]^{\frac{1}{2}} - [2-(\frac{c_R}{c_s})^2] \tag{3.66}$$

and for the damping

$$-\sqrt{2}\,\alpha_R\left([3(\frac{c_p}{c_s})^2 - 1 - 2(\frac{c_R}{c_s})^2] - 2[(\frac{c_p}{c_s})^2 - 1]\right) = (\frac{\omega}{\omega_c})^{\frac{1}{2}}\,\frac{\omega}{c_R}\,\frac{m^2 n}{a}\,[1-(\frac{c_r}{c_s})^2]\,[2-(\frac{c_R}{c_p})^2]\,\frac{c_p}{c_s} \tag{3.67}$$

with

$$m = \frac{\beta K_f}{K_s + \frac{4}{3} G_s + \beta K_f}$$

$$a = \frac{(1-\beta m)\, m}{\beta} \quad \frac{n \rho_f}{(1-n)\rho_s + n\rho_f} \tag{3.68}$$

$$\beta = \frac{1-n}{n}$$

$$\omega_c = \frac{g n}{k}$$

and further more

c_R	velocity surface (Rayleigh) wave
c_p	velocity compressional wave of the first kind
c_s	velocity shear wave
ω	angular frequency

The results obtained by using the equations by Bourbie and by applying the method according to Tajuddin are compared in Table 3.6. From this table it is concluded that the wave velocity calculated by the two models is almost equal. The damping is somewhat lower in Tajuddin's model. However, the damping is proportional with f √f (f: frequency) in both models.

Table 3.6 Comparison results Bourbie and Tajuddin

	Bourbie e.a.		acc. Tajuddin	
frequency	velocity	Im{κ}	velocity	Im{κ}
Hz	m/s	m^{-1}	m/s	m^{-1}
1	52.1	$-0.03*10^{-3}$	52.4	$-0.015*10^{-3}$
10	52.1	$-0.99*10^{-3}$	52.4	$-0.546*10^{-3}$
100	52.1	$-32.2*10^{-3}$	52.3	$-21.9*10^{-3}$

A second check on the solution is be made by comparison with the results of Degrande [1993]. In his thesis he has presented the results of numerical simulations for surface wave propagation. The material properties are shown in Table 3.7. The solid has material damping, so the bulk modulus and the shear modulus of the solid are described by complex variables.

For two frequencies (50 and 100 Hz) the results of the simulations of Degrande and the method following Tajuddin are compared in Table 3.8. The results are in good agreement.

Table 3.7 Material data Mols sand [Degrande, 1992]

property		value	unit
bulkmodulus solid imaginary part	K_s	$298.4*10^6$ $14.9*10^6$	Pa Pa
shearmodulus solid imaginary part	G_s	$111.9*10^6$ $5.6*10^6$	Pa Pa
bulkmodulus fluid	K_f	$2200*10^6$	Pa
volumetric mass solid	ρ_s	2650	kg/m^3
volumetric mass fluid	ρ_f	1000	kg/m^3
apparent volumetric mass	ρ_a	0	kg/m^3
porosity	n	0.388	-
friction coefficient (permeability	b k	$14.8*10^6$ $10*10^{-4}$	$N/(m/s)/m^3$ m/s)

Without material damping (imaginary part of $K_s = 0$ and $G_s = 0$) in the solid, the damping per meter ($1 - e^{-|Im\{\kappa\}|}$) of the surface wave is 0.997 m^{-1} for 50 Hz. This relates to an energy loss which is 10 times as small as with material damping. In this case, the material damping cannot be neglected, if the damping of the surface wave is important.

Motion of solid and fluid

The physical motion of the solid can be calculated from Equation (3.64) and (3.65), using the data in Table 3.5. Only the second solution for φ^s (index 2) and the solution for ψ^s (index 3) are taken into account. The relative motion between the solid and the fluid is small. This can be seen from the values of B_2 and B_3 in Table 3.5. This means the porous medium behaves more or less as a single phase material, see Section 2.2.4.

For a point at the surface ($x_1 = 0$), the motion is calculated. The result is

$$\begin{aligned} \textit{vertical:} \quad & u_1^s(x_1,x_2) = 2.00\, i\, e^{i(\kappa x_2 - \omega t)} \\ \textit{horizontal:} \quad & u_2^s(x_1,x_2) = 0.98\, e^{i(\kappa x_2 - \omega t)} \end{aligned} \tag{3.69}$$

The vertical amplitude is twice the horizontal amplitude. This is about 10% larger than found for an almost incompressible solid in the single phase theory. The phase angle is almost 90°. This is in accordance with the results of single phase theory.

Table 3.8 Comparison celerity and damping
upper part: results Degrande [1992], lower part: method acc. Tajuddin

frequency		50 Hz	100 Hz
	distance (m)	acceleration (m/s^2)	acceleration (m/s^2)
	12.5	0.20	0.155
	25.0	0.14	0.070
	37.5	0.09	0.025
	50.0	0.06	0.005
damping (m^{-1})		0.96	0.91
velocity (m/s)		223	223
frequency		50 Hz	100 Hz
damping (m^{-1})		0.963	0.925
velocity (m/s)		228.1	228.3

Influence of parameters

Since the saturated porous material almost behaves like a single phase material, the dependency of parameters can be derived from the theory of surface waves on a elastic half space. This means that the wave velocity of the Rayleigh wave is mainly influenced by the shear wave velocity and that the damping is mainly influenced by the material damping in the solid.

4 NUMERICAL SOLUTION

4.1 Introduction

In this Chapter a finite element model for the numerical solution of the Biot-Equations is developed. Before the description of the model, a short literature overview is presented in this section. Simon et al. [1986-a, 1986-b] has presented two overview articles, in which several methods are compared for one- and two-dimensional problems. Other useful, more general discussions has been presented by Zienkiewicz & Shiomi [1984] and Ishihara [1970].

Generally, two aspects are of importance for a finite element model that solves a dynamical problem:
- spatial discretisation: defined by the variables which are interpolated in the field (field description)
- temporal discretisation: defined by the method which is used to integrate the spatially discretised equations with respect to time (time integration).

The intermediate result (the equations which are discretised in space, but not in time) are the so-called semi-discretised equations.

An early numerical solution of the Biot-Equations by the finite element method has been presented by Ghaboussi & Wilson [1972]. They have interpolated the displacement of the skeleton and the fluid. This method is called a u-v description. The semi-discretised equations have been integrated with respect to time with several well known time integrators for hyperbolic systems. Sweet [1980] has presented an almost similar method, using the central difference method only, which is explicit.

The u-v description is only valid for a field in which the porosity is constant. In order to simulate non-homogeneous fields, the skeleton displacement and the so-called specific displacement are interpolated. The specific displacement is the relative fluid displacement of the fluid multiplied with the porosity. This method is known as the u-w description. It has been used by e.g. Prevost [1985]. Since the semi-discretised equations are hyperbolic, general time-integration techniques for hyperbolic systems can be applied. Moreover, Prevost has developed an implicit-explicit method. In this method the time step for the fluid and the solid system can be different.

Simon, Wu & Zienkiewicz [1986] have proposed a space discretisation in which the solid displacement and the fluid stress are interpolated. This method is called a u-π description. In order to simplify the equations the relative acceleration of the fluid is neglected. This is reasonable for low frequency and low permeability. This discretisation results in a hyperbolic and a parabolic semi-discretised system, which should be solved simultaneously. This kind of problems can be integrated with respect to time by staggered solutions, which mostly leads to conditional stability [Park, 1983]. For this space discretisation, Zienkiewicz et al. [1988] and Park [1982] have derived a predictor which leads to unconditional stability.

Sandhu, Shaw & Hong [1990] have proposed a space discretisation with three filed variables: They interpolate the solid displacement, specific displacement and fluid stress. This is called a u-π-w description. After condensation of the fluid stress π, the semi-discretised equations are hyperbolic and can be integrated with respect to time by using the Newmark α,β-method.

An other interesting method has been developed by Oka et al. [1994]. Oka's method resembles the u-w description, but the fluid stress is discretised by using finite differences.

Degrande [1992] has presented a finite element method based on the Fourier transform. The problem is solved in the frequency domain. The response on non-harmonic loading must be obtained by inverse Fourier transformation. This implies that the method is limited to linear problems.

Analytical solutions are scarce. An analytical solution has been obtained by Simon e.a. [1984], using Laplace transform. This solution is limited to a dynamic compatible material for one-dimension only. An other (semi-)analytical solution has been obtained by Van der Grinten et al. [1985], using Fourier transform. This solution is limited to linear material behaviour and one-dimension only.

4.2 Development of finite element model

4.2.1 Space discretisation

The field equations (Equation (2.17), Chapter 2) will be discretised by following the standard finite element method. These equations which describe the dynamical behaviour of the saturated porous medium are derived in Chapter 2, and are repeated here for convenience, adding some minor reordering of terms

$$
\begin{aligned}
\rho\, \ddot{u}_j - \sigma^s_{ij,i} &= \pi_{,j} - \rho_f\, \dot{q}_j \\
n\,\pi_{,j} - \frac{b}{n} q_j - (\rho_f + \frac{1}{n}\rho_a)\, \dot{q}_j &= n\rho_f \ddot{u}_j \\
\sigma^s_{ij} &= (K_s - \tfrac{2}{3} G_s)\, u_{k,k}\, \delta_{ij} + G_s\, (u_{i,j} + u_{j,s}) \\
\dot{\pi} &= \frac{1}{n} K_f (\, q_{j,j} + \dot{u}_{j,j}\,)
\end{aligned}
\qquad (4.1)
$$

with $\rho = (1\text{-}n)\rho_s + n\rho_f$.

Equation of motion and constitutive equation for the skeleton

A weak formulation for the equation of motion reads (Galerkin method)

$$\iint_{\Omega} \delta u_j \, [\rho \, \ddot{u}_j - \sigma^s_{ij,i} - \pi_{,j} + \rho_f \dot{q}_j] \, d\Omega = 0 \tag{4.2}$$

Gauss' theorem is applied to the second term

$$-\iint_{\Omega} \delta u_j \, \sigma^s_{ij,i} \, d\Omega = \iint_{\Omega} \delta u_{j,i} \sigma^s_{ij} \, d\Omega - \int_{\Gamma_t} \delta u_j \tilde{\sigma}_{ij} n_i \, d\Gamma \tag{4.3}$$

with n the outward normal at the boundary and the tilde on σ_{ij} refers to the prescribed value of the cartesian stress tensor at the boundary
So, one obtains

$$\iint_{\Omega} \delta u_j \rho \, \ddot{u}_j d\Omega + \iint_{\Omega} \delta u_{j,i} \sigma^s_{ij} \, d\Omega =$$

$$\iint_{\Omega} \delta u_j \pi_{,i} \, \delta_{ij} \, d\Omega - \iint_{\Omega} \delta u_j \, \rho_f \dot{q}_j \, d\Omega + \int_{\Gamma_t} \delta u_j \tilde{t}_j \, d\Gamma \tag{4.4}$$

$$with \quad \tilde{t}_j = n_j \tilde{\sigma}^s_{ij}$$

The second term in Equation (4.4) can be rewritten by using the constitutive equation of the solid (third equation of Equation (4.1))

$$\iint_{\Omega} \delta u_{j,i} \, \sigma^s_{ij} \, d\Omega = \iint_{\Omega} \delta u_{j,i} \, D_{ijkl} \, u_{k,l} \, d\Omega \tag{4.5}$$

$$with \quad D_{ijkl} = (K_s - \tfrac{2}{3} G_s) \, \delta_{ij} \, \delta_{kl} + G_s (\delta_{ik} \, \delta_{jl} + \delta_{il} \, \delta_{jk})$$

D_{ijkl} is the stiffness tensor (a tensor of rank four), see Equation (2.2) Chapter 2.

Equation of motion for the fluid

A weak formulation for the equation of motion for the fluid (Equation (4.1)) reads

$$\iint_{\Omega} \delta q_j \, [\, (\rho_f + \frac{1}{n} \rho_a) \, \dot{q}_j - n \, \pi_{,j} + \frac{b}{n} q_j + n \, \rho_f \ddot{u}_j \,] \, d\Omega = 0 \tag{4.6}$$

Constitutive equation of the fluid

A weak formulation for the constitutive equation of the fluid (Equation (4.1)) reads

$$\iint_{\Omega} \delta\pi \, [\, \dot{\pi} - \frac{1}{n} K_f q_{j,j} - \frac{1}{n} K_f \dot{u}_{j,j} \,] \, d\Omega = 0 \tag{4.7}$$

The second term is rewritten by using Gauss' theorem

$$- \iint_{\Omega} \delta\pi \, \frac{1}{n} K_f \, q_{j,j} \, d\Omega = \iint_{\Omega} \delta\pi_{,j} \frac{1}{n} K_f \, q_j \, d\Omega - \int_{\Gamma_q} \delta\pi \frac{1}{n} K_f \, \tilde{q}_j n_j \, d\Gamma \tag{4.8}$$

Γ_q is the boundary at which the flow through the boundary is prescribed ($q_j = \tilde{q}_j$). Thus, Equation (4.7) becomes

$$\iint_{\Omega} \delta\pi \, \dot{\pi} \, d\Omega + \iint_{\Omega} \delta\pi_{,j} \frac{1}{n} K_f \, q_j \, d\Omega =$$

$$\iint_{\Omega} \delta\pi \, \frac{1}{n} K_f \, \dot{u}_{j,j} \, d\Omega + \int_{\Gamma_q} \delta\pi \, \frac{1}{n} K_f \, \tilde{q}_n \, d\Gamma \tag{4.9}$$

$$with \quad \tilde{q}_n = \tilde{q}_j n_j$$

$\tilde{q}_n$ is the prescribed normal discharge through the boundary.

Discretisation

These weak formulations will be discretised. The index notation used until now, is convenient for three-dimensional problems. However, for implementation of two-dimensional problems (plane strain, axial-symmetry) the matrix notation [e.g. Zienkiewicz and Taylor, 1989] is more convenient. The derivation is limited to two space dimensions. Therefore, the matrix notation will be used here.

The displacement vector is defined and discretised by

$$\boldsymbol{u}(x_1,x_2) = \begin{pmatrix} u_1(x_1,x_2) \\ u_2(x_1,x_2) \end{pmatrix} \approx \boldsymbol{N}_u(x_1,x_2)\, \hat{\boldsymbol{u}} \tag{4.10}$$

with $\mathbf{u}(x_1,x_2)$ the vector with the displacements
$\hat{\mathbf{u}}$ the vector with the numerical approximation of $\mathbf{u}$
$\mathbf{N}_u$ the matrix with the interpolation (or shape) functions

The strain vector is defined by

$$\boldsymbol{\varepsilon}(x_1,x_2) = \begin{pmatrix} \varepsilon_{11}(x_1,x_2) \\ \varepsilon_{22}(x_1,x_2) \\ 2\,\varepsilon_{12}(x_1,x_2) \\ \varepsilon_{33}(x_1,x_2) \end{pmatrix} = \begin{pmatrix} \frac{\partial}{\partial x_1} & 0 \\ 0 & \frac{\partial}{\partial x_2} \\ \frac{\partial}{\partial x_2} & \frac{\partial}{\partial x_1} \\ 0 & 0 \end{pmatrix} \begin{pmatrix} u_1(x_1,x_2) \\ u_2(x_1,x_2) \end{pmatrix} \tag{4.11}$$

$$so \quad \boldsymbol{\varepsilon}(x_1,x_2) = \boldsymbol{D}_u\, \boldsymbol{u}(x_1,x_2) \approx \boldsymbol{D}_u\, \boldsymbol{N}_u(x_1,x_2)\, \hat{\boldsymbol{u}} = \boldsymbol{B}_{uu}(x_1,x_2)\, \hat{\boldsymbol{u}}$$

The stress vector is defined by

$$\boldsymbol{\sigma}(x_1,x_2) = \begin{pmatrix} \sigma_{11}(x_1,x_2) \\ \sigma_{22}(x_1,x_2) \\ \sigma_{12}(x_1,x_2) \\ \sigma_{33}(x_1,x_2) \end{pmatrix} \quad and \quad \boldsymbol{\sigma}(x_1,x_2) = \boldsymbol{S}\,\boldsymbol{\varepsilon}(x_1,x_2) \tag{4.12}$$

$$with \quad \boldsymbol{S} = \begin{pmatrix} K+\frac{4}{3}G & K-\frac{2}{3}G & 0 & K-\frac{2}{3}G \\ K-\frac{2}{3}G & K+\frac{4}{3}G & 0 & K-\frac{2}{3}G \\ 0 & 0 & G & 0 \\ K-\frac{2}{3}G & K-\frac{2}{3}G & 0 & K+\frac{4}{3}G \end{pmatrix} \qquad stiffness\ matrix$$

The relative discharge vector is defined and discretised by

$$\boldsymbol{q}(x_1,x_2) = \begin{pmatrix} q_1(x_1,x_2) \\ q_2(x_1,x_2) \end{pmatrix} \approx \boldsymbol{N}_q(x_1,x_2)\, \hat{\boldsymbol{q}} \tag{4.13}$$

with $\mathbf{q}(x_1,x_2)$ the vector with the relative discharges
$\hat{\mathbf{q}}$ the vector with the numerical approximation of $\mathbf{q}$
$\mathbf{N}_q$ the matrix with the interpolation functions for the relative discharge

The fluid stress is discretised by

$$\pi(x_1,x_2) \approx \boldsymbol{N}_p(x_1,x_2)\, \hat{\boldsymbol{\pi}} \tag{4.14}$$

with $\pi(x_1,x_2)$ the fluid stress
$\hat{\pi}$ the vector with the numerical approximation of π
$\mathbf{N}_p$ the matrix with the interpolation functions for the fluid stress

It is noted that the interpolations functions for the fluid stress are indicated with subscript p and not with subscript π.

Moreover, some derivatives are defined

$$\begin{pmatrix} \frac{\partial}{\partial x_1} \\ \frac{\partial}{\partial x_2} \end{pmatrix} \pi(x_1,x_2) = \boldsymbol{D_p}\ \pi(x_1,x_2) \approx \boldsymbol{D_p}\ \boldsymbol{N_p}(x_1,x_2)\ \hat{\boldsymbol{\pi}} = \boldsymbol{B_{pp}}(x_1,x_2)\ \hat{\boldsymbol{\pi}} \tag{4.15}$$

$$\frac{\partial u_1(x_1,x_2)}{\partial x_1} + \frac{\partial u_2(x_1,x_2)}{\partial x_2} = {}^t\boldsymbol{D_p}\ \boldsymbol{u}(x_1,x_2) \approx$$
$$ {}^t\boldsymbol{D_p}\ \boldsymbol{N_u}(x_1,x_2)\ \hat{\boldsymbol{u}} = \boldsymbol{B_{pu}}(x_1,x_2)\ \hat{\boldsymbol{u}} \tag{4.16}$$

The vectors $\hat{\mathbf{u}}$, $\hat{\mathbf{q}}$ and $\hat{\pi}$ are the vectors with (nodal) unknowns. These vectors still depends on time. The matrices $\mathbf{N}_u$, $\mathbf{N}_q$ and $\mathbf{N}_p$ are the matrices with interpolation functions. The matrices $\mathbf{B}_{uu}$, $\mathbf{B}_{pp}$ and $\mathbf{B}_{pu}$ are the matrices with derivatives of the shape functions. These matrices depends on the space coordinates. From now on, this dependency will not be written any more.

The variations can be treated in the same way as the field variables. For example, the variation of the displacement $\delta\mathbf{u}$ can be written as

$$\delta\,\boldsymbol{u} \approx \boldsymbol{N_u}\ \delta\hat{\boldsymbol{u}} \tag{4.17}$$

The summation convention over an subscript is written as a vector product in the vector notation.

This discretisation method leads to the following equations.

For the solid (from Equation(4.4)):

$$\iint_\Omega {}^t\boldsymbol{N_u}\ \rho\ \boldsymbol{N_u}\ d\Omega\ \frac{d^2\hat{\boldsymbol{u}}}{dt^2} + \iint_\Omega {}^t\boldsymbol{B_{uu}}\ \boldsymbol{S}\ \boldsymbol{B_{uu}}\ d\Omega\ \hat{\boldsymbol{u}} =$$
$$\iint_\Omega {}^t\boldsymbol{N_u}\ \boldsymbol{B_{pp}}\ d\Omega\ \hat{\pi} - \iint_\Omega {}^t\boldsymbol{N_u}\ \rho_f\ \boldsymbol{N_q}\ d\Omega\ \frac{d\hat{\boldsymbol{q}}}{dt} + \int_{\Gamma_t} {}^t\boldsymbol{N_u}\ \tilde{t}\ d\Gamma \tag{4.18}$$

For the equation of motion of the fluid (from Equation (4.6)):

$$\iint_\Omega {}^t\boldsymbol{N_q}\ (\rho_f + \frac{1}{n}\rho_a)\ \boldsymbol{N_q}\ d\Omega\ \frac{d\hat{\boldsymbol{q}}}{dt} - \iint_\Omega {}^t\boldsymbol{N_q}\, n\, \boldsymbol{B_{pp}}\ d\Omega\ \hat{\pi}$$
$$+ \iint_\Omega {}^t\boldsymbol{N_q}\ \frac{b}{n}\ \boldsymbol{N_q}\ d\Omega\ \hat{\boldsymbol{q}} = -\iint_\Omega {}^t\boldsymbol{N_q}\ n\rho_f\ \boldsymbol{N_u}\ d\Omega\ \frac{d^2\hat{\boldsymbol{u}}}{dt^2} \tag{4.19}$$

For the constitutive equation of the fluid (from Equation (4.9)):

$$\iint_\Omega {}^t\boldsymbol{N}_p \, \boldsymbol{N}_p \, d\Omega \, \frac{d\hat{\pi}}{dt} + \iint_\Omega {}^t\boldsymbol{B}_{pp} \, \frac{1}{n} \, K_f \, \boldsymbol{N}_q \, d\Omega \, \hat{\boldsymbol{q}} = \iint_\Omega {}^t\boldsymbol{N}_p \, \frac{1}{n} \, K_f \, \boldsymbol{B}_{pu} \, d\Omega \, \frac{d\hat{\boldsymbol{u}}}{dt} + \int_{\Gamma_q} {}^t\boldsymbol{N}_p \, \frac{1}{n} \, K_f \, \tilde{\boldsymbol{q}}_n \, d\Gamma \tag{4.20}$$

Symbolic equations

The equations (4.18), (4.19) and (4.20) can be written in symbolic form.

For the solid

$$\boldsymbol{M}_{uu} \frac{d^2\hat{\boldsymbol{u}}}{dt^2} + \boldsymbol{K}_{uu}\hat{\boldsymbol{u}} = \boldsymbol{K}_{up}\hat{\pi} + \boldsymbol{C}_{uq}\frac{d\hat{\boldsymbol{q}}}{dt} + \boldsymbol{f}_u \tag{4.21}$$

with

$$\boldsymbol{M}_{uu} = \iint_\Omega {}^t\boldsymbol{N}_u \rho \boldsymbol{N}_u \, d\Omega \tag{4.22-a}$$

$$\boldsymbol{K}_{uu} = \iint_\Omega {}^t\boldsymbol{B}_{uu} \, \boldsymbol{S} \, \boldsymbol{B}_{uu} \, d\Omega \tag{4.22-b}$$

$$\boldsymbol{K}_{up} = \iint_\Omega {}^t\boldsymbol{N}_u \, \boldsymbol{B}_{pp} \, d\Omega \tag{4.22-c}$$

$$\boldsymbol{C}_{uq} = -\iint_\Omega {}^t\boldsymbol{N}_u \, \rho_f \boldsymbol{N}_q \, d\Omega \tag{4.22-d}$$

$$\boldsymbol{f}_u = \int_{\Gamma_t} {}^t\boldsymbol{N}_u \tilde{\boldsymbol{t}} \, d\Gamma \tag{4.22-e}$$

For the equation of motion of the fluid

$$\boldsymbol{C}_{qq} \frac{d\hat{\boldsymbol{q}}}{dt} + \boldsymbol{K}_{qp}\hat{\pi} + \boldsymbol{K}_{qq}\hat{\boldsymbol{q}} = \boldsymbol{M}_{qu}\frac{d^2\hat{\boldsymbol{u}}}{dt^2} \tag{4.23}$$

with

$$\boldsymbol{C}_{qq} = \iint_\Omega {}^t\boldsymbol{N}_q \, (\rho_f + \frac{1}{n}\rho_a) \, \boldsymbol{N}_q \, d\Omega \tag{4.24-a}$$

$$\boldsymbol{K}_{qp} = -\iint_\Omega {}^t\boldsymbol{N}_q n \, \boldsymbol{B}_{pp} \, d\Omega \tag{4.24-b}$$

$$\boldsymbol{K}_{qq} = \iint_\Omega {}^t\boldsymbol{N}_q \, \frac{b}{n} \, \boldsymbol{N}_q \, d\Omega \tag{4.24-c}$$

$$M_{qu} = - \iint\limits_{\Omega} {}^{t}N_q \; n \; \rho_f \, N_u \; d\Omega \tag{4.24-d}$$

For the constitutive equation of the fluid

$$C_{pp} \frac{d\hat{\pi}}{dt} + K_{pq}\hat{q} = C_{pu}\frac{d\hat{u}}{dt} + h_p \tag{4.25}$$

with

$$C_{pp} = \iint\limits_{\Omega} {}^{t}N_p \; N_p \; d\Omega \tag{4.26-a}$$

$$K_{pq} = \iint\limits_{\Omega} {}^{t}B_{pp} \; \frac{1}{n} \; K_f \, N_q \; d\Omega \tag{4.26-b}$$

$$C_{pu} = \iint\limits_{\Omega} {}^{t}N_p \; \frac{1}{n} \; K_f \, B_{pu} \; d\Omega \tag{4.26-c}$$

$$h_p = \int\limits_{\Gamma_q} {}^{t}N_p \; \frac{1}{n} \; K_f \, \tilde{q} \; d\Gamma \tag{4.26-d}$$

4.2.2 Choice of interpolations

In this paragraph the interpolation functions are described. Again, the discussion is limited to two space dimensions. Both triangular elements with three and with six nodes are considered, see Figure Figure 4.1.

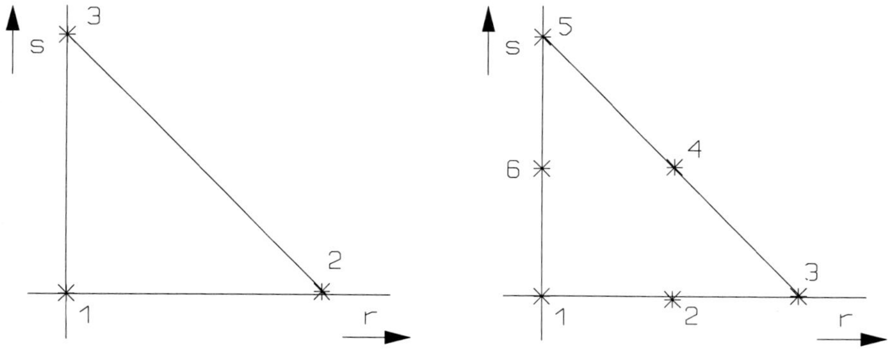

Figure 4.1 Reference element (local coordinates)

Three interpolation functions must be defined for:

- the displacements of skeleton
- the fluid stress
- the relative discharge of the fluid.

The displacements in r-direction and s-direction are interpolated by linear functions on the element with three nodes and quadratic functions on the element with six nodes.

$$u_r(r,s) = \sum_{j=1}^{M_{nod}} N_u^j(r,s)\; u_r^j$$
$$u_s(r,s) = \sum_{j=1}^{M_{nod}} N_u^j(r,s)\; u_s^j \tag{4.27}$$

with M_{nod} = 3 for linear elements and M_{nod} = 6 for the quadratic element.
Isoparametric elements are considered. Therefore the element with three nodes can be called linear and the element with six nodes can be called quadratic.

Changing to the vector form gives

$$\boldsymbol{u}(r,s) = \boldsymbol{N}_u(r,s)\;\hat{\boldsymbol{u}} \tag{4.28}$$

with

$$\hat{\boldsymbol{u}} = {}^t(u_r^1\; u_s^1\; u_r^2\; u_s^2\; u_r^3\; u_s^3\; ..\,)$$
$$\textit{and}$$
$$\boldsymbol{N}_u = \begin{pmatrix} N_u^1 & 0 & N_u^2 & 0 & N_u^3 & 0 & .. \\ 0 & N_u^1 & 0 & N_u^2 & 0 & N_u^3 & .. \end{pmatrix} \tag{4.29}$$

The interpolation of the fluid stress and the relative discharge needs some special attention. Due to the discretisation, the fluid stress is the so-called constraint variable and the relative discharge is the so-called primary variable [e.g. Zienkiewicz and Taylor, 1989]. The interpolation of the relative discharge and the fluid stress must obey Babuska-Brezzi criterion. Therefore, The relative discharge can be interpolated one degree lower than the fluid stress.

The fluid stress can be interpolated by linear functions on the linear element and by quadratic functions on the quadratic element.

$$\pi(r,s) = \sum_{j=1}^{M_{nod}} N_p^j(r,s)\; \pi^j \tag{4.30}$$

with in this case

$$N_p^i(r,s) = N_u^i(r,s) \tag{4.31}$$

Changing to the vector form gives

$$\pi(r,s) = N_p(r,s)\,\hat{\pi} \tag{4.32}$$

with

$$\hat{\pi} = {}^t(\ \pi^1\ \pi^2\ \pi^3\ ..\ ) \quad \textit{and} \quad N_p = \left(\ N_p^1\ N_p^2\ N_p^3\ ..\ \right) \tag{4.33}$$

The relative discharge might be discontinuous on an interface between two different materials. If an interface coincides with an element boundary only continuity in perpendicular direction is required. If continuity of the parallel discharge is also imposed, so-called exaggerated continuity occurs.

To prevent the occurrence of exaggerated continuity two methods of interpolation are possible:

- a standard polynomial interpolation, releasing the continuity at the boundaries
- a Raviart-Thomas interpolation

Using a polynomial interpolation offers the possibility of a jump of the discharge parallel with the boundary, but also a jump of the discharge perpendicular to the boundary. In a Raviart-Thomas interpolation only the perpendicular discharge is interpolated, so continuity of the discharge perpendicular to the boundary is ensured. From this point of view the Raviart-Thomas interpolation is preferable. However, the polynomial interpolation is easier to implement. This advantage increases for quadratic and cubic elements.

In the polynomial interpolation the relative discharge is constant over the linear element (defined in the bari-centre of the element) and linear on the quadratic element (defined in the corner nodes of the element).

In the Raviart Thomas interpolation, only the component perpendicular on the edges is interpolated [e.g. Chavent & Roberts, 1991]. This interpolation is successfully applied on groundwater flow problems by Kaasschieter and Huijben [1992]. The unknowns are defined in the edge nodes of the element. The Raviart-Thomas interpolation is tested for rectangular elements with four nodes [Hölscher & Barends, 1993]. In that case the interpolation leads to an unstable solution for the present dynamical system. The Raviart-Thomas interpolation is not tested for triangular elements. The theory is shown in appendix 4A.

For the interpolation with polynomials the interpolation formula reads

$$q_r(r,s) = \sum_{j=1}^{M_{Q\text{-}nod}} N_q^j(r,s)\, \hat{q}^j$$
$$q_s(r,s) = \sum_{j=1}^{M_{Q\text{-}nod}} N_q^j(r,s)\, \hat{q}^j \tag{4.34}$$

with $M_{Q\text{-nod}} = 1$ for the linear element and $M_{Q\text{-nod}} = 3$ for the quadratic element.

Changing to vector notation gives

$$\boldsymbol{q}(r,s) = \boldsymbol{N}_q(r,s)\, \hat{\boldsymbol{q}} \tag{4.35}$$

with

$$\hat{\boldsymbol{q}} = {}^t(q_s^1\ q_r^1\ q_r^2\ q_s^2\ q_r^3\ q_s^3\ ..)$$
$$\boldsymbol{N}_q = \begin{pmatrix} N_q^1 & 0 & N_q^2 & 0 & N_q^3 & 0 & .. \\ 0 & N_q^1 & 0 & N_q^2 & 0 & N_q^3 & .. \end{pmatrix} \tag{4.36}$$

From now on the superimposed ^ on the numerical approximation is omitted, since no confusion between the physical variable and its numerical approximation can occur any more.

4.2.3 Time discretisation

General overview

The three systems defined by the Equations (4.21), (4.23) and (4.25) are coupled. The time integration method is presented in this section.
The two systems for the fluid are considered as one system and the system for the solid displacements are considered as a second system.

The first system (π,q) is a mixed formulation. The relative discharge is eliminated from the system on element level. The remaining first order system for the pore pressure is solved by Euler implicit. This remaining system is called the fluid system.

The second system for the displacements (u) is solved by Hilbert-Hughes-Taylor α-method. This system is called the solid system.

Both systems have terms in the right-hand-side, which depends on the variables of the other system. These are treated explicitly. For these terms predictors are used, accordingly to the method of Heun. The advantage of the method of Heun over the method of Euler is the gain of one order in time: Euler is first order, Heun is second order.

Elimination of discharge

First the (semi-discretised) equation (4.23) for the derivative of the relative discharge is considered:

$$C_{qq} \frac{dq}{dt} + K_{qp}\, \pi + K_{qq}\, q = M_{qu} \frac{d^2 u}{dt^2} \tag{4.37}$$

This equation is solved in time, according to

$$C_{qq} \left(\frac{dq}{dt}\right)^{n+1} + K_{qp}\, \pi^{n+1} + K_{qq}\, q^{n+1} = M_{qu} \left(\frac{d^2 u}{dt^2}\right)^{n+1} \tag{4.38}$$

The superscript n relates to the previous time level, the superscript n+1 to the current time level. The relative discharge at the current time level can be estimated by using Euler implicit. From the Taylor series, one finds

$$q^{n+1} = q^{n} + \Delta \left(\frac{dq}{dt}\right)^{n+1} \tag{4.39}$$

Δ is the time step. So, the derivative of q at the current time level is approximated by using

$$\left(\frac{dq}{dt}\right)^{n+1} = \frac{q^{n+1}-q^{n}}{\Delta} \tag{4.40}$$

The new value of the discharge q^{n+1} can be derived explicitly from the original equation

$$C_{qq}\, q^{n+1} + \Delta\, K_{qq}\, q^{n+1} = \Delta\, M_{qu} \left(\frac{d^2 u}{dt^2}\right)^{n+1} + C_{qq}\, q^{n} - \Delta\, K_{qp}\, \pi^{n+1} \tag{4.41}$$

so, if the inverse matrix exists,

$$q^{n+1} = (C_{qq} + \Delta\, K_{qq})^{-1} \left[\Delta\, M_{qu} \left(\frac{d^2 u}{dt^2}\right)^{n+1} + C_{qq}\, q^{n} - \Delta\, K_{qp}\, \pi^{n+1} \right] \tag{4.42}$$

Secondly the (semi-discretised) equation (4.25) for the derivative of the pore-pressure is considered, with a zero prescribed relative discharge on the boundary Γ_q

$$C_{pp} \frac{d\pi}{dt} + K_{pq}\, q = C_{pu} \frac{du}{dt} \tag{4.43}$$

This is solved in time, according to

$$C_{pp} \left(\frac{d\pi}{dt}\right)^{n+1} + K_{pq}\, q^{n+1} = C_{pu} \left(\frac{du}{dt}\right)^{n+1} \tag{4.44}$$

Using Equation (4.42) for q^{n+1} one finds

$$C_{pp}\left(\frac{dp}{dt}\right)^{n+1} - \Delta\, K_{pq}\, Q_{qq}\, K_{qp}\, p^{n+1} =$$

$$C_{pu}\left(\frac{du}{dt}\right)^{n+1} - K_{pq}\, Q_{qq}\,[C_{qq}\, q^{n} + \Delta\, M_{qu}\left(\frac{d^2u}{dt^2}\right)^{n+1}] \tag{4.45}$$

or

$$C_{pp}\left(\frac{d\pi}{dt}\right)^{n+1} + K_{pp}\, \pi^{n+1} = f_p^{n+1} \tag{4.46}$$

with

$$Q_{qq} = (C_{qq} + \Delta\, K_{qq})^{-1}$$

$$K_{pp} = -\Delta\, K_{pq}\, Q_{qq}\, K_{qp} \tag{4.47}$$

$$f_p^{n+1} = C_{pu}\left(\frac{du}{dt}\right)^{n+1} - K_{pq}\, Q_{qq}\,[C_{qq}\, q^{n} + \Delta\, M_{qu}\left(\frac{d^2u}{dt^2}\right)^{n+1}]$$

Time integration of system for fluid stress

Now the system (Equations (4.46))

$$C_{pp}\left(\frac{d\pi}{dt}\right)^{n+1} + K_{pp}\, \pi^{n+1} = f_p^{n+1} \tag{4.48}$$

must be solved. Again the Euler implicit method is applied.

Using

$$\left(\frac{d\pi}{dt}\right)^{n+1} = \frac{\pi^{n+1}-\pi^{n}}{\Delta} \tag{4.49}$$

one finds

$$\frac{1}{\Delta}\, C_{pp}\, \pi^{n+1} + K_{pp}\, \pi^{n+1} = f_p^{n+1} + \frac{1}{\Delta}\, C_{pp}\, \pi^{n} \tag{4.50}$$

or

$$S_{pp}\, \pi^{n+1} = g_p^{n+1} \tag{4.51}$$

with

$$S_{pp} = \frac{1}{\Delta} C_{pp} + K_{pp}$$
$$g_p^{n+1} = \frac{1}{\Delta} C_{pp}\, \pi^n + f_p^{n+1} \tag{4.52}$$

Equation (4.52), together with Equation (4.42), are referred to as the fluid system (first system) and are solved numerically.

Time integration of the equation for the solid

The equation for the solid reads (cf. Equation (4.21))

$$M_{uu}\ddot{u} + K_{uu}u = K_{up}\pi + C_{uq}\dot{q} + f_u \tag{4.53}$$

This Equation is solved by Hilbert-Hughes-Taylor α-method [Hilbert, Hughes & Taylor, 1977]. The displacement, velocity and acceleration at time $t = t^{n+\alpha}$ are approximated by using u^n, v^n, a^n, u^{n+1}, v^{n+1} and a^{n+1} respectively. The time-discretised equation then reads

$$M_{uu}a^{n+1} + (1+\alpha)\, K_{uu}u^{n+1} - \alpha\, K_{uu}\, u^n =$$
$$K_{up}\, \pi^{n+\alpha} + C_{uq}\left(\frac{dq}{dt}\right)^{n+\alpha} + f_u(t^{\,n+\alpha}) \tag{4.54}$$

with $t^{n+\alpha} = (1+\alpha)\, t^{n+1} - \alpha\, t^n = t^{n+1} + \alpha\, \Delta$.
The superscript $n+\alpha$ means the variable evaluated at $t = t^{n+\alpha}$.

In order to obtain a stable solution the value of α is limited to $-1/3 < \alpha < 0$. The displacements and velocities are calculated by using

$$u^{n+1} = u^n + \Delta v^n + \frac{1}{2}\Delta^2\, [\, (1-2\beta)\, a^n + 2\beta\, a^{n+1}\,]$$
$$v^{n+1} = v^n + \Delta[\, (1-\gamma)\, a^n + \gamma a^{n+1}\,] \tag{4.55}$$

with $\beta = 0.25\, (1-\alpha)^2$ and $\gamma = 0.5\, (1-2\alpha)^2$.

Two predictors $\tilde{u}$ and $\tilde{v}$ for the displacement and the velocity respectively, are defined by

$$\tilde{u}^{n+1} = u^n + \Delta v^n + \frac{1}{2}\Delta^2\, (1-2\beta)\, a^n$$
$$\tilde{v}^{n+1} = v^n + \Delta\, (1-\gamma)\, a^n \tag{4.56}$$

From the first part (for the displacement) of the Equations (4.55) and (4.56) the acceleration a^{n+1} can be calculated

$$a^{n+1} = \frac{u^{n+1} - \tilde{u}^{n+1}}{\beta\, \Delta^2} \tag{4.57}$$

The final system can be calculated by elimination of a^{n+1} from the time-discretised equation (4.54), giving

$$[\frac{1}{\beta} M_{uu} + (1+\alpha)\,\Delta^2 K_{uu}]\, u^{n+1} = \alpha\,\Delta^2 K_{uu}\, u^n + \frac{1}{\beta} M_{uu}\, \tilde{u}^{n+1} + \Delta^2\,[f_u(t^{n+\alpha}) + K_{up}\,\pi^{n+\alpha} + C_{uq}\left(\frac{dq}{dt}\right)^{n+\alpha}] \qquad (4.58)$$

This equation can be written symbolically as

$$S_{uu}\, u^{n+1} = g_u^{n+1} \qquad (4.59)$$

with

$$\begin{aligned} S_{uu} &= \frac{1}{\beta} M_{uu} + (1+\alpha)\,\Delta^2 K_{uu} \\ g_u^{n+1} &= \alpha\,\Delta^2 K_{uu}\, u^n + \frac{1}{\beta} M_{uu}\, \tilde{u}^{n+1} + \Delta^2\,[f_u(t^{n+\alpha}) + K_{up}\,\pi^{n+\alpha} + C_{uq}\left(\frac{dq}{dt}\right)^{n+\alpha}] \end{aligned} \qquad (4.60)$$

Solution of the coupled system

The two matrix equations ((4.51) and (4.59)) must be solved simultaneously, but for calculating the right hand side of each system the result from the other system is needed. This can be solved by applying the method of Heun [Van Kan, 1992]. The method of Heun requires one iteration each time step. Therefore the following equations are solved in the order as shown.

$$\begin{aligned} S_{pp}\, \check{\pi}^{n+1} &= g_p^{n+1}(t, q^n, q^{n-1}, v^n, a^n) \\ S_{uu}\, \check{u}^{n+1} &= g_u^{n+1}(t, u^n, v^n, a^n, \check{\pi}^{n+1}, \check{q}^{n+1}, q^n) \\ S_{pp}\, \pi^{n+1} &= g_p^{n+1}(t, q^n, q^{n-1}, \check{v}^{n+1}, \check{a}^{n+1}) \\ S_{uu}\, u^{n+1} &= g_u^{n+1}(t, u^n, v^n, a^n, \pi^{n+1}, q^{n+1}, q^n) \end{aligned} \qquad (4.61)$$

The first two steps are the so-called predictor steps. The variables solved are predicted values, written with a superimposed ˇ . The last two steps are the so-called corrector steps. After each step, the discharge is calculated from the fluid stress by using Equation (4.42), the acceleration and velocity are calculated from the displacement by using Equation (4.55).

Convergence

In order to obtain convergence to the exact solution the numerical scheme must be consistent and stable. Here, the convergence is not proven by a complete mathematical treatment, but it is shown that the convergence is plausible.

The solid system is solved according to a well-known method in literature, therefore the consistency of this system is assured.

The fluid system is solved using a mixed method together with dynamic condensation. The mixed method used in the finite element method is treated by Zienkiewicz and Taylor [1989] for static problems. Based on a simplified Babuska-Brezzi criterium, they show the convergence of the p-q formulation for the quadratic element. The criterium they use is satisfied for a patch of elements, although it is violated for a single linear element. Therefore, consistency is assured.

The second question is related to the stability of the time integration which is used. The method employed (Euler implicit) is unconditionally stable if the matrix K_{pp} is symmetric and definite positive. This is indeed the case: since C_{qq} and K_{qq} are symmetric and definite positive matrices, the matrix $C_{qq} + \Delta K_{qq}$ is symmetric and definite positive ($\Delta > 0$). Thus the inverse Q_{qq} exists and it is symmetric and definite positive. Since the matrices K_{pq} and K_{qp} are the transposed of each other (the relation $K_{pq} = \gamma\, {}^{T}K_{qp}$ holds, with γ a constant), the matrix $K_{pp} = K_{pq}\, Q_{qq}\, K_{qp}$ is also symmetric and definite positive. Therefore, the method of Euler implicit is unconditionally stable for this system.

The method of Heun is conditionally stable. Therefore, it is concluded that the applied method is conditionally stable and consistent, thus convergence is assured if the timestep is within the stability region of the method of Heun.

During the testing of the finite element scheme, the stability condition turned out to be related to a critical timestep. This critical timestep is defined as the time the wave of the first kind needs to travel the shortest distance between two nodes.

4.2.4 Conditions at boundaries and interfaces and treatment of dry porous material and impermeable material

This section discusses the implementation of the conditions at boundaries of a saturated material. Furthermore it discusses the way the impermeable material and the dry porous material are treated in the numerical model, including the interfaces.

In Chapter 2 three types of interfaces are discussed in general terms:

- interface between to saturated porous materials
- saturated porous material to impermeable material
- phreatic surface inside a porous material

The numerical treatment of these boundary conditions are discussed shortly in this section. Due to the choice of field variables the case of an interface between two saturated porous materials is automatically treated correctly by imposing continuity on u^s and π along the interface and is not discussed in this section.

A special problem is the occurrence of artificial boundary conditions in the finite element mesh. The treatment of these boundaries is discussed in this section too.

Ordinary boundary conditions

The solid of the saturated porous medium is treated as a conventional continuum. Thus the displacement at the boundaries are the essential boundary conditions and the stresses at the boundaries are the natural boundary conditions.

For the fluid of the saturated porous medium the fluid stress is the constrained variable, so at the boundary the fluid stress is the essential boundary condition. A prescribed relative discharge is a natural boundary condition. In a dynamical problem a prescribed relative discharge is physically not very relevant and, therefore, not implemented.

Impermeable material and dry porous material

In the finite element model the impermeable solid and the dry porous material are treated in the same way as the solid of the saturated porous medium. This coincides with the well-known treatment in literature for a single phase material. The resulting stiffness matrices are stored in the system matrix of the solid system. Thus, the vector with unknowns of the solid system contains displacements of the skeleton of the saturated porous material, the skeleton of the dry porous material and the impermeable solid. The system matrix of the solid system refers to the skeleton of the saturated porous material, the skeleton of the dry porous material and the impermeable solid. The vector with unknowns of the fluid system contains fluid stresses in the saturated porous material only. The system matrix of the fluid system refers to the fluid in the saturated porous material only and doesn't exist on the domain of the impermeable material and the dry porous material.

Interface between a saturated porous medium and an impermeable medium

The required interface conditions are derived by Deresiewicz and Skalak [1963], see chapter 2. Limited to the two-dimensional case these conditions read

$$\begin{gathered} u_n^i = u_n , \quad u_t^i = u_t , \quad q_n = 0 \\ \sigma_n^i = \sigma_n + \pi , \quad \sigma_t^i = \sigma_t \end{gathered} \tag{4.62}$$

with u displacement of solid
σ solid stress
π fluid stress
q relative discharge
superscript i refers to the impermeable solid
subscript n: normal to interface, t: along interface

The interface is a natural boundary condition for the fluid system, i.e. the normal relative discharge q_n is zero. The displacements of impermeable solid and the skeleton of the saturated porous material are equal at the interface, due to the fact that both displacements are stored in the same vector (the vector of unknowns of the solid system). However, this treatment doesn't satisfy the fourth condition of Equation (4.62). Therefore, the fluid stress at the interface must be imposed as an additional stress on the impermeable solid. This is a natural boundary condition for the solid system (stress is prescribed).

Phreatic surface

Limited to the two-dimensional case, the conditions at a phreatic surface read

$$\begin{gathered} u_n^d = u_n \,, \quad u_t^d = u_t \,, \\ \sigma_n^d = \sigma_n \,, \quad \sigma_t^d = \sigma_t \,, \quad \pi = 0 \end{gathered} \tag{4.63}$$

with u displacement of solid
σ solid stress
π fluid stress
superscript d refers to the skeleton of the dry solid
subscript n: normal to interface, t: along interface

The influence of capillary forces is neglected.

A phreatic surface is an essential boundary condition for the fluid system, defined by the condition that the fluid stress is zero ($\pi = 0$). The solid of the dry porous material above the phreatic surface is modelled as a single phase material, and the unknowns are stored in the vector of unknowns of the solid system. Therefore, the displacements of the skeleton of the dry material and the saturated material are equal. Since the fluid stress is zero at the phreatic surface, no additional stress must be imposed.

Transmitting boundary

Artificial boundaries are introduced by the fact that the dimensions of the finite element mesh are limited, but in reality the extension of the domain is infinite. After some time reflections of these artificial boundaries arrive in the region of interest. This is not realistic.

A simple solution for this quandary can be obtained by creating a very large mesh, but this causes considerable calculation time. In the finite element model Titan, simple dashpots are used. For monochromatic waves in a single phase material, these boundary conditions are sufficient. For shock loading on a two-dimensional domain, the efficiency is not very high.

The equations for the transmitting boundary are derived using one-dimensional wave propagation. The fluid system and the solid system are treated separately.

For a wave propagating in positive x-direction in the fluid the discharge and the fluid stress are related by

$$\pi = - Z_f \, q_x \tag{4.64}$$

with $Z_f = 1/n \sqrt{(K_f \rho_f)}$ the mechanical impedance.

This equation can be used to eliminate the normal discharge in the boundary integral of Equation (4.20). This procedure leads to an additional term in the stiffness matrix K_{pp} of Equation (4.46), based on the integral

$$\int_{\Gamma_q} \delta\pi \, \frac{K_f}{n \, Z_f} \, \pi \, d\Gamma \tag{4.65}$$

In symbolic notation this matrix is

$$\boldsymbol{K}_{pp} = \int_{\Gamma_{tb}} \boldsymbol{N}_p^t \, \frac{K_f}{n \, Z_f} \, \boldsymbol{N}_p \, d\Gamma \tag{4.66}$$

This matrix must be lumped.

For the transmitting boundary of the solid system, a similar method is applied. For a compressional and a shear wave propagating in positive direction the solid velocities and the stresses are related by

$$\begin{aligned} \sigma_n &= - Z_n \, v_n \\ \sigma_t &= - Z_s \, v_t \end{aligned} \tag{4.67}$$

with $Z_n = \sqrt{((K_s + 4\,G_s/3)/\rho_{eq})}$ the compressional impedance and $Z_s = \sqrt{(G_s/\rho_{eq})}$ the shear impedance.

This equation can be used to eliminate the normal traction in the boundary integral of Equation (4.4). This procedure leads to a damping matrix in Equation (4.21), in symbolic notation

$$\begin{gathered} \boldsymbol{C}_{uu} = \int_{\Gamma_{tb}} \boldsymbol{N}_u^t \, \boldsymbol{D}_{tb} \, \boldsymbol{N}_u \, d\Gamma \\ \textit{with} \\ \boldsymbol{D}_{tb} = \begin{pmatrix} n_1^2 Z_n + n_2^2 Z_s & n_1 n_2 (Z_n - Z_s) \\ n_1 n_2 (Z_n - Z_s) & n_2^2 Z_n + n_1^2 Z_s \end{pmatrix} \end{gathered} \tag{4.68}$$

Here, n_1 and n_2 are the component of the outward normal vector on the boundary. This matrix must be lumped to a tri-diagonal matrix.

4.3 Non-linear model for skeleton

4.3.1 Introduction

Generally the constitutive behaviour of the solid skeleton is not linear. A non-linear model is implemented in the finite element program Titan. The non-linear model is treated in this section.

The behaviour of a solid (e.g. sand) which is saturated with a fluid (e.g. water) is not simple to describe. In the description herein, emphasis is placed on two phenomena:

- firstly the fact that a saturated sample can liquefy during undrained cyclic shear loading due to pore water pressure build-up
- secondly the fact that the stiffness of a sample decreases significantly if the pre-consolidation stress is exceeded.

The model contains a consolidation surface (f_c) and an elasto-plastic concept including non-linear hardening. The elasto-plastic model with non-linear kinematic hardening is defined by a yield surface (f_η), a plastic potential (g_η) and a hardening relation. Moreover, a shear-failure criterion is stated. The equations defining the model are treated in this section.

4.3.2 Elastic model and consolidation surface

First the behaviour of a sample during an oedometer test is considered. During the oedometer test the stiffness strongly depends on the stress level. Under the so called over-consolidation stress, the sample is relatively stiff, above the consolidation stress the sample is relatively weak. This behaviour is described using a bi-linear model, in which the stiffness is controlled by the position of the stress state in the stress space relatively to the consolidation surface.

The consolidation surface which controls the elastic stiffness of the material, is defined by

$$f_c = (\eta_{ij}\,\eta_{ij})^{\frac{1}{2}} + M_m \ln\left(\frac{\sigma_m}{\sigma_{m,\,max}}\right) = 0 \qquad (4.69)$$

with η_{ij} the stress ratio tensor ($\eta_{ij} = (\sigma_{ij} - \sigma_m\,\delta_{ij})/\sigma_m$ and $\sigma_m = \sigma_{ii}/3$)
M_m material constant
$\sigma_{m,max}$ the maximum value of σ_m on the surface.

The value of $\sigma_{m,max}$ is updated if $f_c > 0$ in order to hold $f_c = 0$. At the consolidation surface the material is normally consolidated and the elastic stiffness is determined by using the compression index. Within the surface the material is overconsolidated and the elastic stiffness is determined by using the swelling index.

The elastic moduli are defined by

$$K_s = \frac{(1+e_o)\ \sigma_m}{S_i}$$
$$G_s = \frac{3(1-2\nu)}{2(1+\nu)}\ \frac{(1+e_o)\ \sigma_m}{S_i} \tag{4.70}$$

with ν Poison ratio
e_o initial void ratio
σ_m mean stress
S_i the swelling index κ if $f_c < 0$ and the compression index λ if $f_c = 0$

The Poison ratio is chosen constant.

For a triaxial test the consolidation surface is drawn in Figure 4.2 together with the other surfaces contained in the model.

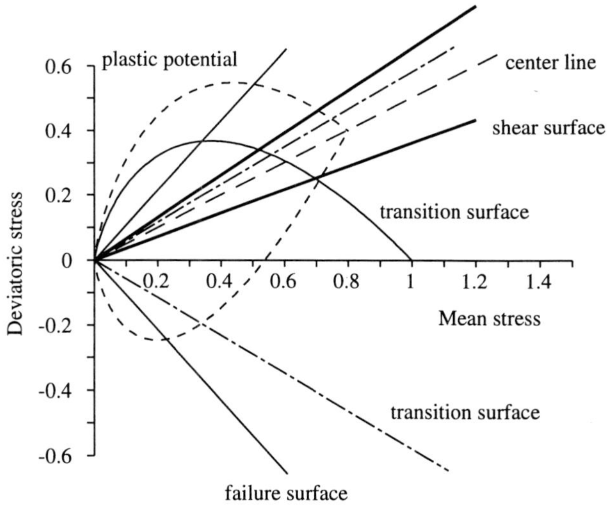

Figure 4.2 Surfaces used in non-linear model

4.3.3 Elasto-plastic model

The elasto-plastic model is a modified version of the model proposed by Oka et al. [1992]. During a undrained triaxial test the stiffness of the sample is strongly stress dependant and large volumetric changes may occur. The model derived by Oka et al. is specially meant for this kind of loading and is able to describe liquefaction of a sample during an undrained test.

Yield surface

The yield surface is defined by

$$f_\eta = [(\eta_{ij} - x_{ij})(\eta_{ij} - x_{ij})]^{\frac{1}{2}} - R_\eta = 0 \tag{4.71}$$

with x_{ij} kinematic hardening tensor for shear plasticity (initial value = 0)
R_η a scalar constant, which defines the elastic region

In the stress invariants space (deviator stress on mean stress, in this section called p-q space), this surface is defined by two straight lines. The bisection of these two lines is defined by the hardening tensor. In the π-plane the surface is a circle.

Non-linear hardening rule

The increment of the hardening tensor is defined by

$$dx_{ij} = B_\eta (A_\eta \, de^p_{ij} - x_{ij} \, d\gamma^p) \tag{4.72}$$

with A_η and B_η material dependant parameters
A_η is related to the maximum compression at failure
B_η is related to the rate of hardening
de^p_{ij} the incremental deviatoric plastic strain tensor

$$de^p_{ij} = d\varepsilon^p_{ij} - \frac{1}{3} d\varepsilon^p_{kk} \, \delta_{ij} \tag{4.73}$$

$d\varepsilon^p_{ij}$ the increment of the plastic strain tensor
$d\gamma^p$ the second invariant of the incremental plastic strain tensor

$$d\gamma^p = (de^p_{ij} \, de^p_{ij})^{\frac{1}{2}} \tag{4.74}$$

This hardening rule leads to a decreasing hardening when the plastic strain increases.

Plastic potential

The plastic potential is defined by

$$g_\eta = [(\eta_{ij} - x_{ij})(\eta_{ij} - x_{ij})]^{\frac{1}{2}} - M^* \ln\left(\frac{\sigma_m}{\sigma_{ma}}\right) = 0 \tag{4.75}$$

with M^* a constant which still depends on the transition state (see below):
outside the surface ($f_t > 0$) $M^* = M_m$
inside the surface ($f_t < 0$) $M^* = -\eta / \ln(\sigma_m / \sigma_{mc})$ with $0 < M^* < M_m$$_*$
Once the normally consolidated region has been reached the constant M^* is always equal to M_m.
σ_{ma} a scaling factor
The stress σ_{mc} equals the mean consolidation stress $\sigma_{m,max}$.
$\eta = \sqrt{(\eta_{ij} \, \eta_{ij})}$

The volumetric changes during plasticity are controlled by the plastic potential. The currently chosen plastic potential leads inside the transition surface to dilatancy which increases with the stress ratio. Outside the transition surface the sample dilates strongly for small values of the mean stress and contracts for high values of the mean stress. If σ_{ma} and σ_{mc} are equal, the model is associated inside the transition surface.

Transition surface

The transition surface is defined by

$$f_t = [(\eta_{ij} - \eta_{ij}^o)\,(\eta_{ij} - \eta_{ij}^o)]^{\frac{1}{2}} - M_m = 0 \tag{4.76}$$

with the superscript o referring to the variable at the end of consolidation. In the p-q space this surface is a straight line.

Shear failure

The sample fails on shear failure when the shear failure surface is reached. The shear failure surface is defined by

$$f_{failure} = (\eta_{ij}\,\eta_{ij})^{\frac{1}{2}} - D_{f_\eta}\, M_m = 0 \tag{4.77}$$

with $D_{f\eta} \geq 1$, a material dependent factor which shows that the shear strength is reached after crossing the transition surface. This factor can also be used to take into account that the strength during dynamic loading might be higher then during static loading. In the p-q space this is a straight line, which is steeper than the transition surface (with $\eta^o = 0$).

4.3.4 Implementation

The elasto-plastic stiffness tensor

The elasto-plastic stiffness tensor is derived from the partition of the strain in a elastic and a plastic part and the consistency equation [e.g. Zienkiewicz & Taylor 1989].

The partition of the strain gives

$$\begin{aligned} d\sigma_{kl} &= D_{klij}\, d\varepsilon_{ij}^e \\ &= D_{klij}\,(d\varepsilon_{ij} - d\varepsilon_{ij}^p) \end{aligned} \tag{4.78}$$

The consistency equation reads

$$df_\eta = \frac{\partial f_\eta}{\partial \sigma_{ij}}\, d\sigma_{ij} + \frac{\partial f_\eta}{\partial x_{ij}}\, dx_{ij} = 0 \tag{4.79}$$

Using the hardening rule (Equation (4.72)) and writing the plastic strain as

$$d\varepsilon^{p}_{ij} = \Lambda \frac{\partial g_{\eta}}{\partial \sigma_{ij}} \tag{4.80}$$

with Λ the plastic multiplier, the Equations (4.78) and (4.79) are rewritten

$$\begin{aligned} d\sigma_{kl} + D_{klij} \frac{\partial g_{\eta}}{\partial \sigma_{ij}} \Lambda &= D_{klij}\, d\varepsilon_{ij} \\ \frac{\partial f_{\eta}}{\partial \sigma_{ij}} d\sigma_{ij} + \frac{\partial f_{\eta}}{\partial x_{ij}} H_{ij}\, \Lambda &= 0 \end{aligned} \tag{4.81}$$

with

$$H_{ij} = B \left[A \frac{\partial g_{\eta}}{\partial s_{ij}} - x_{ij} \left(\frac{\partial g_{\eta}}{\partial s_{kl}} \frac{\partial g_{\eta}}{\partial s_{kl}} \right)^{\frac{1}{2}} \right] \tag{4.82}$$

with s_{ij} the deviator stress tensor.

The plastic multiplier can be eliminated. The result is the elasto-plastic stiffness tensor S_{mnij}, defined from

$$\begin{aligned} d\sigma_{mn} &= \left(D_{mnij} - \frac{1}{H} D_{mnpq} \frac{\partial g_{\eta}}{\partial \sigma_{pq}} \frac{\partial f_{\eta}}{\partial \sigma_{kl}} D_{klij} \right) d\varepsilon_{ij} \\ &\quad or \\ d\sigma_{mn} &= S_{mnij}\, d\varepsilon_{ij} \end{aligned} \tag{4.83}$$

with

$$H = \frac{\partial f_{\eta}}{\partial \sigma_{mn}} D_{mnij} \frac{\partial g_{\eta}}{\partial \sigma_{ij}} - \frac{\partial f_{\eta}}{\partial x_{ij}} H_{ij} \tag{4.84}$$

Integration of the non-linear model

Once the increment of the strain tensor $d\varepsilon_{ij}$ is known, the increment of the stress tensor $d\sigma_{ij}$ can be calculated from (4.78), by taking into account that the stress must be at the yield surface in case of plasticity. These requirements can be met by using an iterative method. The integration method is shown in Table 4.1. Since the yield function f_{η} is smooth and the method of Newton-Raphson is applied, the convergence of the iteration method is assured.

The initial value of Λ is calculated by choosing Λ equal to zero. This means that the elastic stress increment is calculated, which is an over estimation of the stress increment [c.f. Brinkgreve, 1994]. The integration of the stress tensor and the hardening tensor is carried out by a numerical integration method, e.g. the classical Runge-Kutta method.

Table 4.1 Flowchart for iterative solution of elasto-plastic law

$$\textit{while} \; (\, | f^k | \geq \delta \,)$$

$$\Lambda^{k+1} = \Lambda^k - \frac{f^k}{(df^k / d\lambda)}$$

$$d\varepsilon_{ij}^p = \Lambda^{k+1} \frac{\partial g_\eta^p}{\partial \sigma_{ij}}$$

$$d\sigma_{mn} = D_{mnij} \, (\, d\varepsilon_{ij} - d\varepsilon_{ij}^p \,)$$

$$\sigma_{mn}^{k+1} = \sigma_{ij}^{k} + d\sigma_{mn}$$

$$\eta_{ij}^{k+1} = \frac{\sigma_{ij}^{k+1} - \sigma_{mean}^{k+1} \, \delta_{ij}}{\sigma_{mean}^{k+1}}$$

$$dx_{ij} = B \, (\, A \; de_{ij}^p - x_{ij}^k \, d\gamma^p \,)$$

$$x_{ij}^{k+1} = x_{ij}^{k} + dx_{ij}$$

$$f^{k+1} = [\, (\, \eta_{ij}^{k+1} - x_{ij}^{k+1} \,) \, (\, \eta_{ij}^{k+1} - x_{ij}^{k+1} \,) \,]^{\frac{1}{2}} - R_\eta$$

$$k = k+1$$

$$\textit{end while loop}$$

k: iteration number
δ: required accuracy

4.3.5 Results of simulation for model tests

In this Section the possibilities of the model are elucidated by considering an undrained triaxial test and a drained oedometer test during cyclic loading. Table 4.2 shows the material properties together with additional test data. The influence of some important parameters is discussed.

Table 4.2 Material data for studying behaviour of material model

property	value	unit
e_0 initial void ratio	1.286	-
λ compression index	0.030	-
κ swelling index	0.006	
ν Poison ratio	0.33	-
M_m stress ratio at shear failure	0.77	-
R_η radius elastic region	0.14	-
B_η parameter hardening rule	600.0	-
A_η parameter hardening rule	1.48	-
σ_{ma} uniting stress factor	100.0	kPa
$D_{f\eta}$ (dynamic) correction factor	4.0	-
τ_{min} minimum stress (*)	-33 / .01	kPa
τ_{max} maximum stress (*)	33 / 33	kPa
$\sigma_{m,o}$ initial mean stress	100	kPa
$\sigma_{m,p}$ mean pre-consolidation stress	120	kPa
$\Delta\varepsilon$ strain increment (*)	.05 / .01	10^{-3}

(*) first number for triaxial test, second for oedometer test

The following tests are considered in this Section:

1 drained oedometer test
2 undrained triaxial test
3 undrained triaxial test with $D_{f\eta}$ = 1.35
4 undrained triaxial test on an over-consolidated sample
5 undrained triaxial test with A_η = 1.00
6 drained oedometer test with large unloading amplitude

For convenience of the reader the picture are shown by using soil mechanical stress definition: compressional stress is positive.

Figure 4.3 shows the mean stress in the sample depending on vertical strain during the oedometer test. The change from over-consolidation to normally consolidation during the first loading is clearly seen. The following loading cycles are fully elastic.

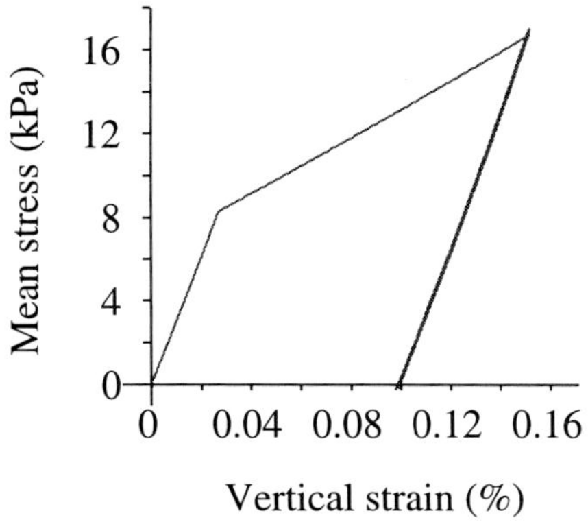

Figure 4.3 Simulation of oedometer test
Data Table 4.2

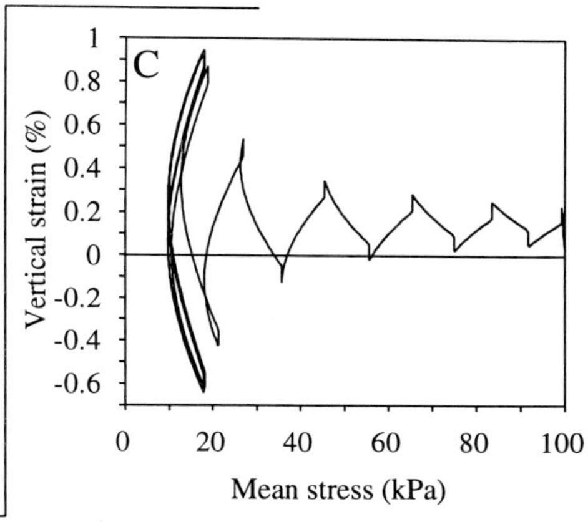

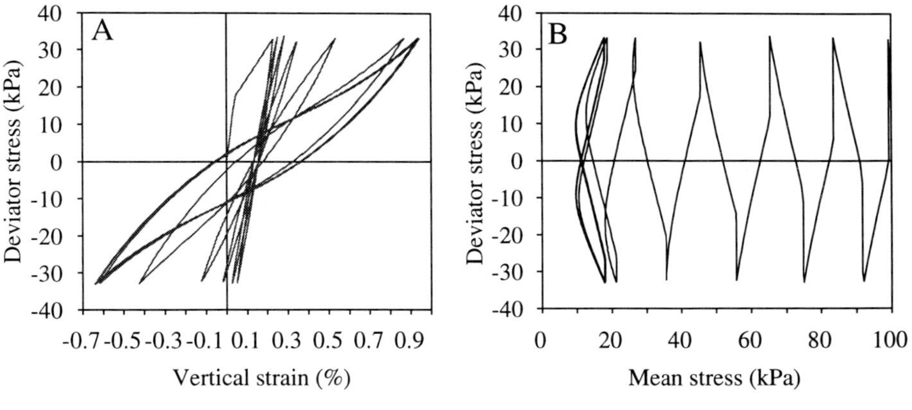

Figure 4.4 Simulation of undrained triaxial test
Data Table 4.2

Figure 4.4 shows the behaviour of the sample during undrained triaxial loading. During the cyclic loading the stiffness of the sample decreases, due to the decreasing mean stress. Directly after stress reversal the sample reacts elastic, as can be seen from the constant value for the mean stress (vertical line) in the q-p space. After six cycles, a stable cyclic behaviour is reached. The results of cyclic tests [e.g. Ishihara & Okada, 1978] show a similar behaviour.

Figure 4.5 shows the behaviour of the sample during undrained triaxial loading. The material properties are equal to those in the previous example, however, the (dynamic) correction factor $D_{f\eta}$ = 1.35. During the cyclic loading the stiffness of the sample decreases, due to the decreasing mean stress. Now, the sample doesn't reach a stable cyclic behaviour, but it fails during the third cycle. Due to the shear failure the sample looses its stiffness completely.

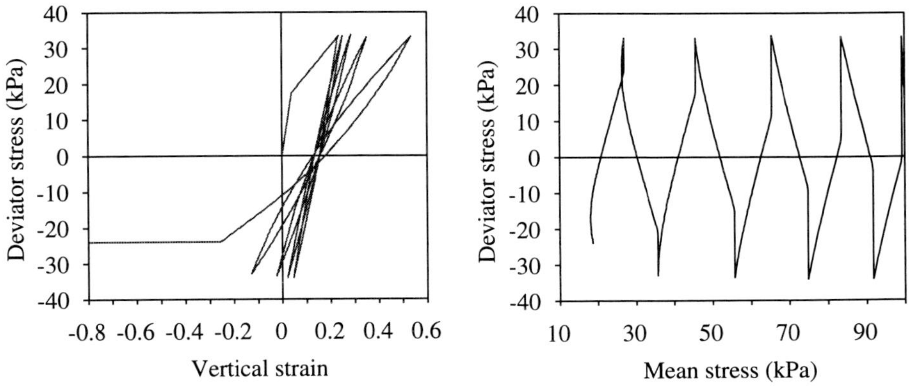

Figure 4.5 *Simulation of triaxial test with shear failure*
Data Table 4.2, but $D_{f\eta}$ = 1.35

The influence of the over-consolidation state is studied for the case with the pre-consolidation stress increased to $\sigma_{m,o}$ = 300 kPa (over-consolidation ratio of 3.0). Over-consolidation also influences the value of the constant M_m, the elastic region R_η, initial void ratio e_o, the swelling index κ and the compression index λ. Figure 4.6 give the results of a simulation with M_m = 0.51, R_η = 0.21, e_o = 1.0, κ = 0.007 and λ = 0.035. In order to prevent for shear failure, the (dynamic) correction factor $D_{f\eta}$ = 4.0 is used. The result must be compared with the results in Figure 4.4.

The over-consolidated sample with needs almost 20 cycles before reaching the stable cyclic behaviour. The transition surface is reached after 11 cycles, but this cannot be seen in Figure 4.6. The vertical strain during stable cyclic behaviour is smaller than for the normal consolidated sample.

Figure 4.7 shows the influence of the hardening parameter A_η which is decreased to 2/3 of the original value. Comparing Figure 4.7 and Figure 4.4, the area of the loop in the deviator stress on vertical strain diagram has changed, thus, the sample dissipates more energy per cycle. Since the rate of hardening depends of the hardening parameter A_η, the mean stress decreases faster than with the original value of A_η.

Finally two oedometer tests are considered. They resemble the one discussed above, however, in the first one the maximum load is increased (so that the stress path is outside the yield surface at the compressional side) and in the second one the minimum load is

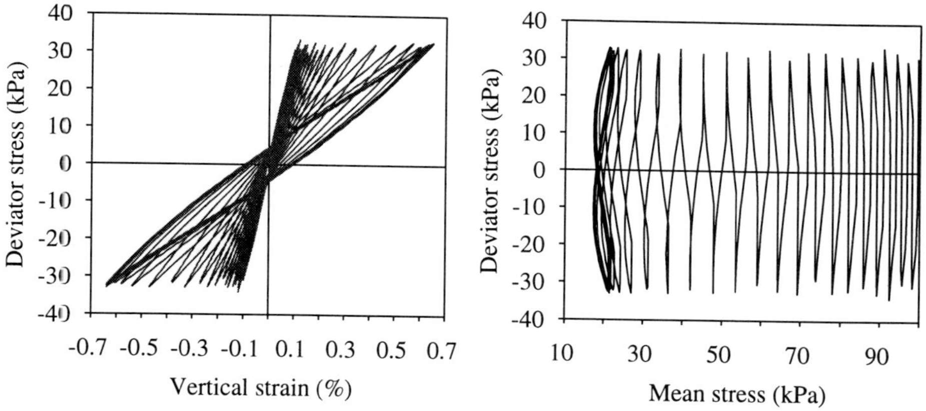

Figure 4.6 *Simulation of triaxial test for over-consolidated soil* Data Table 4.2, but $\sigma_{m,o}$ = 300 kPa, M_m = 0.51, R_η = 0.21, e_o = 1.0, λ = 0.035, κ = 0.007

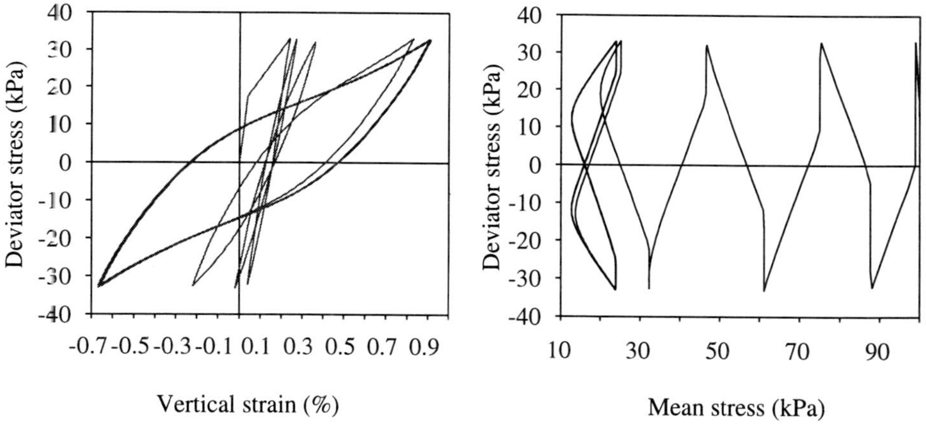

Figure 4.7 *Simulation of triaxial test with other hardening rule*
Data Table 4.2, but A_η = 1.0

increased as well (so that the stress path is outside the yield surface at the extensional side).

Figure 4.8 shows the results of the simulations. For large loading amplitudes (left graph), the results are good. For large unloading, so-called over-ratchetting occurs. For simulation of material behaviour at high mean stress level combined with large loading and unloading, the model seems less satisfactory.

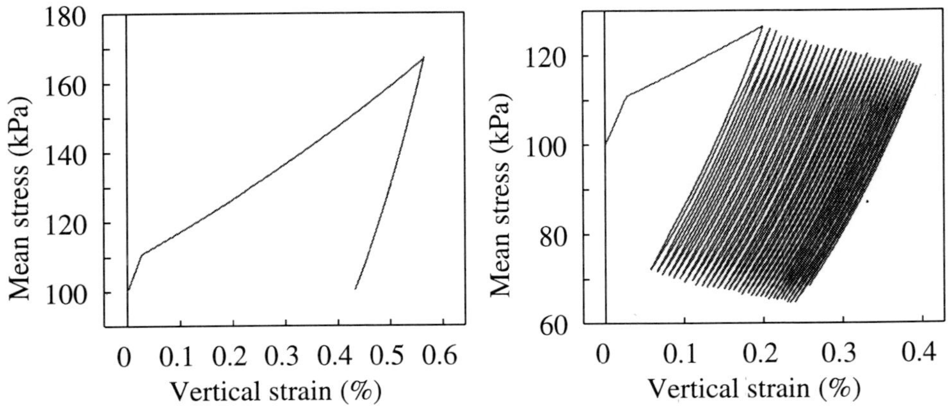

Figure 4.8 *Simulation of oedometer tests with very large loading (a) and large unloading (b), Data Table 4.2 but τ_{min} and/or τ_{max} changed*

4.4 Test examples: one-dimensional compressional waves

Introduction

In this section several examples of the use of Titan for one-dimensional problems are shown. The target of this section is the validation of the program Titan. Therefore, the discussion is limited to one-dimensional problems, since comparison with analytical solutions is possible. The simulation of two-dimensional problems are discussed in chapter 6.

The following examples are treated:

- A compressible fluid in a frictionless stiff skeleton
- Simulation of the test of Van der Grinten
- Efficiency of the transmitting boundary
- The application of the interface elements
- The consolidation of a sand layer after a sudden unloading
- The application of the bi-linear elastic model

The fluid system

The solution method for the fluid system is a newly developed integration scheme. Its' behaviour is shortly examined in this section.

For this examination the deformation of the solid is chosen zero ($u^s_j = 0$) and the friction and the apparent mass are also chosen zero ($b = 0$ and $\rho_a = 0$). A one-dimensional problem is solved. In this case, the analytical solution shows a wave which propagates

Table 4.3 Material data for test fluid system

property	symbol	value	unit
bulk modulus	K_f	2000	MPa
volumetric mass	ρ_f	1000	kg/m^3
apparent mass	ρ_a	0	kg/m^3
porosity	n	0.4	-

undistorted through the material. At one boundary the fluid is loaded by a pressure load with a rise time of 300 μs and constant afterwards. At the other boundary two types of boundary conditions are tested: an impermeable end (q = 0) and a transmitting boundary. The length of the domain is 0.5 m. Table 4.3 shows the material data. The celerity ($c = \sqrt{K_f/\rho_f}$) is 1.41 m/ms and the impedance ($Z = 1/n \sqrt{(K_f \rho_f)}$) is $3.54*10^6$ Ns/m.

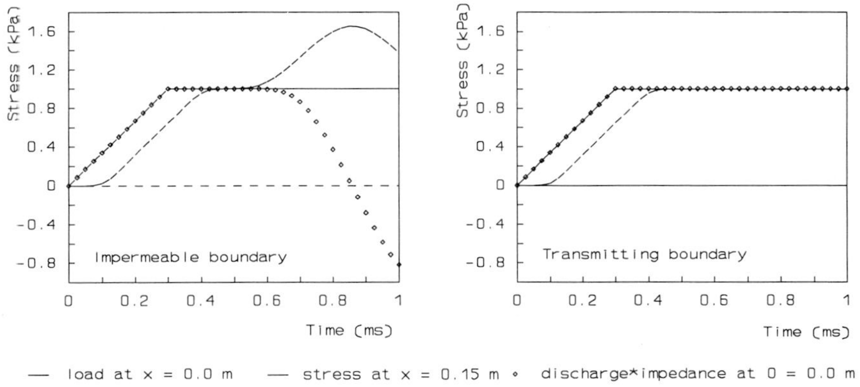

Figure 4.9 Results test integration scheme for the fluid

Figure 4.9 shows the results of the calculations. The left-hand-side shows the case with the impermeable boundary and the right-hand-side shows the case with the transmitting boundary. The applied fluid stress at the loaded boundary, the calculated discharge (divided by the impedance Z of the material) at the loaded boundary and the fluid stress at 0.15 m from the loaded boundary are shown in both cases.

From the left figure it is concluded that the integration method results in the expected analytical behaviour of a single phase material. The sharp kinks in the loading are distorted, due to the numerical dispersion which is generated by the integration scheme. After t = 0.6 ms, the influence of the reflections of the impermeable end, can be clearly observed.

The result with the transmitting boundary, shown in the right-hand-side of Figure 4.9, doesn't show the influence of these reflections. So, in the one-dimensional case without friction, the transmitting boundary is very efficient.

Simulation of test of Van der Grinten

Van der Grinten [1985] has carried out a shock-tube test, see Figure 4.10. The shock-tube exists of a low pressure chamber and a high pressure chamber. The sample of saturated sandstone has been placed in the low pressure chamber. Suddenly, the separation between the high-pressure chamber and low-pressure chamber is removed. A pressure wave propagates in the water in the low-pressure chamber and loads the sample of saturated sandstone. In the sample the pore water pressure and effective stress are measured.

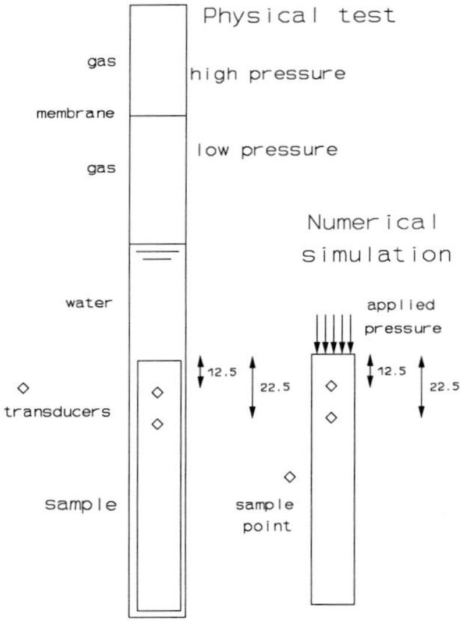

Figure 4.10 Shock tube test of Van der Grinten [1985]

The material data are shown in Table 4.3 [Van der Grinten, 1987]. The loading of the saturated porous sandstone exists of a step load with a rise time of 40 μs.
For testing the finite element program, the problem is discretised with a single row of triangular element with size 25 * 25 mm^2.

The effective stress and pore fluid stress are calculated at 125 and 225 mm from the loaded side of the saturated porous medium. The results of the finite element program are compared with analytical result. The analytical results are based on Fourier transform of the problem, see Chapter 3 [c.f. Smeulders, 1993].

Figure 4.11 shows the numerical results and the analytical results for the situation without apparent mass. Figure 4.12 shows the same problem with an apparent mass of 400 kg/m^3. From these Figures it is concluded that the results of the numerical program are in good agreement with the analytical results. The wave of the second kind shows more dispersion. The final value of the stresses in good agreement.

Efficiency of transmitting boundary for saturated porous medium

The efficiency of the transmitting boundary is evaluated by comparing the simulation of the test of Van der Grinten with the analytical solution. The cases with a transmitting boundary and without a transmitting boundary are considered. The material data are similar as in the preceding section, see Table 4.4. The length of the mesh is half of the length used in the preceding section.

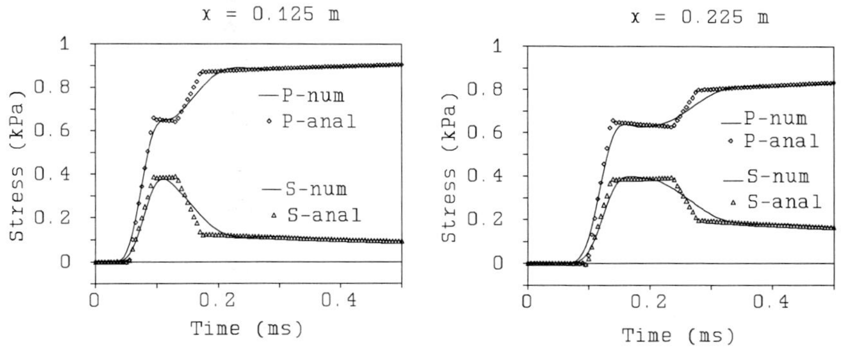

Figure 4.11 *Comparison numerical and analytical results, no apparent mass*

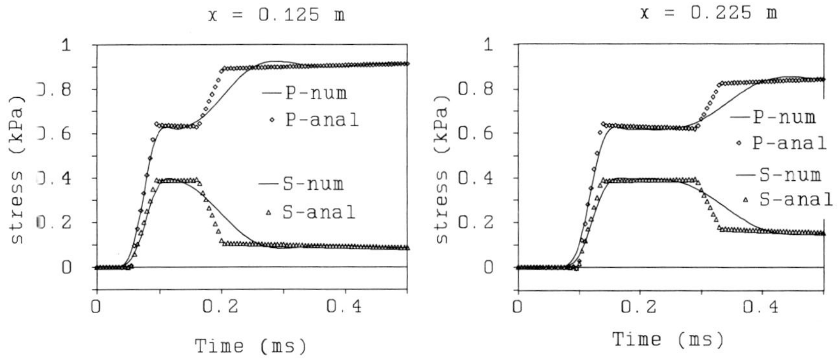

Figure 4.12 *Comparison numerical and analytical results apparent mass = 400 kg/m^3*

Figure 4.13 shows the fluid stress and the solid stress at x = 0.225 m for the three cases. The solution with the transmitting boundary and the solution without the transmitting boundary are equal until the reflections of the lower boundary arrive at t = 0.35 ms. Then the simulation with the transmitting boundary is much closer to the analytical solution than the case without the transmitting boundary.
The case with the transmitting boundary is not identical to the analytical solution. This might be due to two effects:

- the transmitting element is not perfect, since it takes not into account the fact that the solid and the fluid are coupled.
- the reflections of the wave of the second kind are included in the reflections.

The first point is an unavoidable disadvantage of the usage of simple dashpots. The influence of the wave of the second kind can be eliminated by placing the transmitting boundary further away, so that the wave of the second kind is damped.

Table 4.4 Material data for simulation test of Van der Grinten [1987]

property	symbol	value	unit
bulk modulus solid	K_s	2790	MPa
shear modulus solid	G_s	1286	MPa
bulk modulus fluid	K_f	2000	MPa
volumetric mass solid	ρ_s	2600	kg/m^3
volumetric mass fluid	ρ_f	1000	kg/m^3
apparent mass	ρ_a	0 / 400	kg/m^3
porosity	n	0.3	-
friction permeability	b k	$1.6*10^6$ $0.55*10^{-3}$	kg/sm^3 m/s

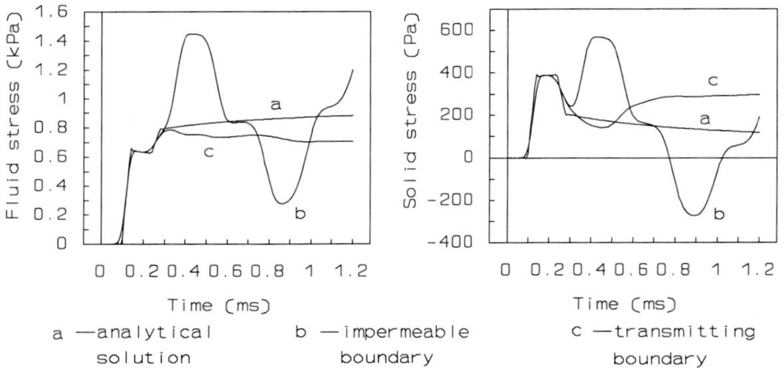

Figure 4.13 Efficiency of transmitting boundary, simulation test of Van der Grinten

<u>Interface behaviour</u>

In order to test the formulation of the interface behaviour, a one-dimensional problem with an interface is studied. The problem which is solved analytical in chapter 3, is solved numerically again. The material data are presented in table 3.3 in chapter 3. For the solid skeleton and the impermeable solid a Poison ratio of $\nu = 0.3$ is used. The length of the impermeable block is 2 m, the load rises in 1 ms to the final value of 1 kN/m^2.

Figure 4.14 compares the numerically and analytically calculated solid stress at 0.1 m, 0.2 m and 0.4 m from the interface. The results are in good agreement.

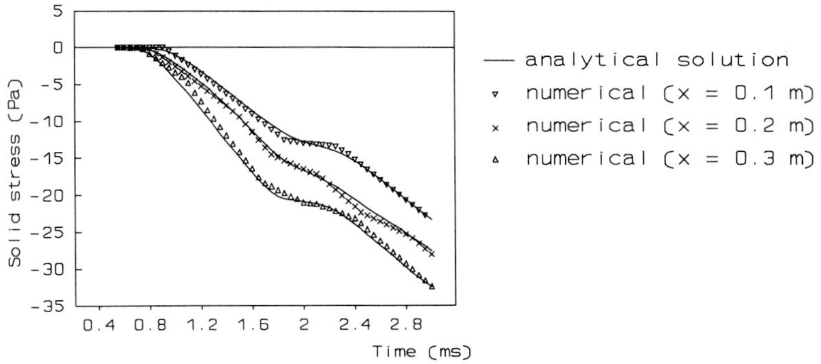

Figure 4.14 Comparison numerical and analytical solution close to interface

Consolidation

In order to check whether the model can calculate consolidation, the consolidation behaviour of the test of Van der Grinten is calculated.

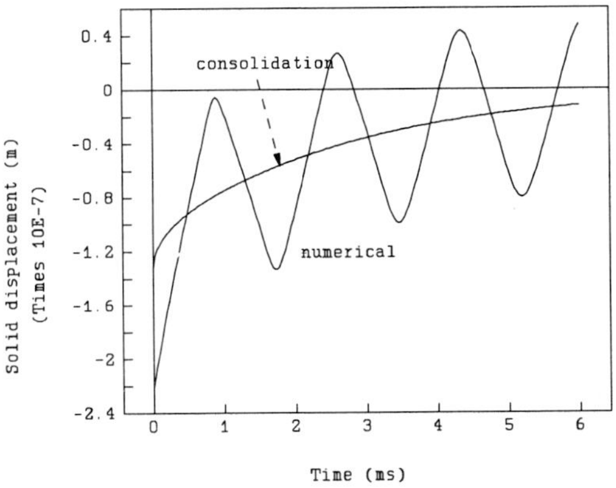

Figure 4.15 Displacement of loaded surface compared with value based on consolidation theory

Figure 4.15 shows the displacement of the loaded surface depending on time calculated with Titan together with the value based on quasi-static consolidation theory. From this figure it is clear that in this linear case, the dynamic wave propagation and the consolidation can be superimposed.

Figure 4.16 shows the calculated the relative discharge at x = 0.0 m, x = 0.125 m and x = 0.225 m from the loaded end. During the beginning the arrival of the first wave (positive value) and the second wave (negative value) can be seen. The relatives discharge can be divided into two parts: a quasi-static smooth curve and the multiple reflections superimposed on the smooth consolidation curve. The smooth curve starts after the arrival of the second wave. In the calculated relative discharge the positive discharge is related to the wave of the first kind, the negative values with the wave of the second kind.

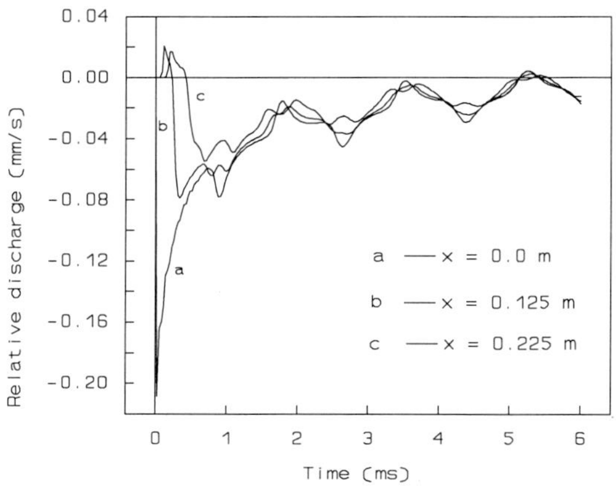

Figure 4.16 Specific discharge consolidation test

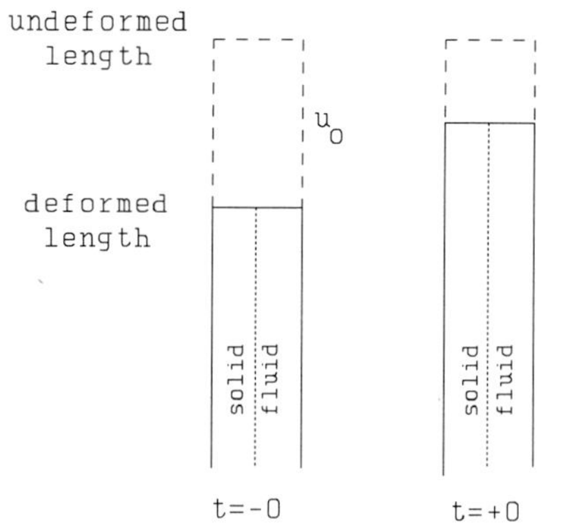

Figure 4.17 Deformations directly before and after unloading

It is remarkable that the displacement related to the consolidation at start is not equal to the total initial displacement (0.22 μm). This can be explained by the following reasoning: Firstly, the solid and fluid move simultaneously until equilibrium is reached. This displacement is called the instantaneous displacement. Secondly, the consolidation starts from this equilibrium. Now the fluid flows through the solid. The situation is explained in Figure 4.17.

By considering the porous column as two elastic bars (one for the fluid and one for the solid), the instantaneous displacement can be calculated from the equilibrium. The force in the solid is equal to the force in the fluid

$$K_s\,(u_0 - \delta u) = \frac{1}{n} K_f\,\delta u \quad \Rightarrow \quad \delta u = \frac{K_s\,u_0}{K_s + \frac{1}{n} K_f}$$

with u_0 static displacement
δu displacement difference
K_s constrained modulus of the solid
K_f bulk modulus of the fluid
n porosity

Using the material data from Table 4.4, one finds for the instantaneous displacement δu = 0.403. This means that the consolidation starts at (1.0-0.403) 0.22 = 0.131 μm. This value is used for the calculation of the static consolidation.

Non-linear model, bi-linear elasticity

In the last section of this chapter the usage of the non-linear model in the finite element program Titan is shown for a one-dimensional problem. The example relates to the bi-linear elastic model. An over-consolidated dry material is loaded by a step-load at one end. The pre-consolidation stress is reached during dynamic loading. Table 4.5 shows the relevant material data for this example.

Table 4.5 Material data for example wave propagation in bi-linear elastic model

variable	symbol	value	unit
volumetric mass	ρ_s	1882	kg/m^3
initial void ratio	e_o	1.286	-
compression index	λ	0.03	-
swelling index	κ	0.006	-
Poison ratio	ν	0.33	-
pre-consolidation stress	$\sigma_{m,max}$	-1500	kPa
initial stress	$\sigma_{m,ini}$	-1000	kPa

Figure 4.18 shows the dynamic compressional stress along the bar at several moments. The bi-linear behaviour can be clearly seen. The over-consolidation stress is -1500.0 kPa, which is 1.5 times the initial stress of -1000.0 kPa. Below the pre-consolidation stress, the material is relatively stiff. The wave propagates with a high velocity. However, above the pre-consolidation level, the material is much weaker, so the wave propagates with a lower velocity. The change in propagation velocity takes place at the pre-consolidation stress

level, thus at -1500 kPa. Based on the values of the compression index and the swelling index, the high velocity is 2.2 times the low velocity. This value is also found in Figure 4.18.

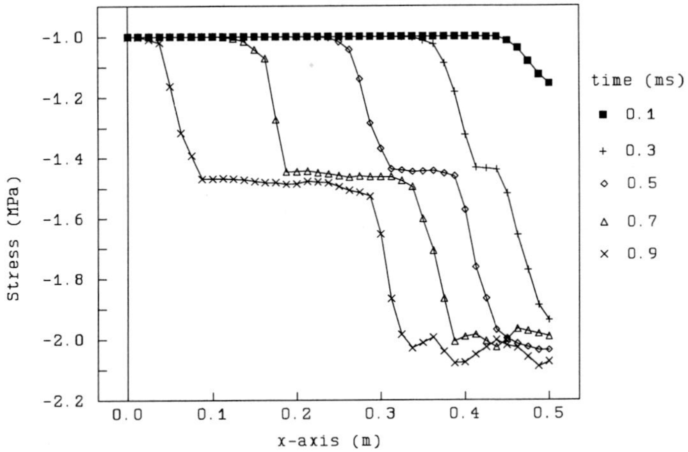

Figure 4.18 One-dimensional wave propagation in bi-linear elastic model

5 DYNAMIC EXPERIMENTS

5.1 Introduction

Three dynamic experiments were carried out. These experiments point at the behaviour of saturated porous media. The methods and results are described in this chapter.

The first test is about a pile-driving simulator. The main target of this test is the creation of a laboratorium set-up to study the differences between the wave radiation under a pile toe in dry sand and saturated sand. By a new method of indirect optical measurement accurate information on the wave radiation is obtained.

The second test is a full scale in-situ test. During the driving and testing of a pile the pore water pressure and accelerations in the soil at 18 m below the surface are measured. The acceleration measurements provide a complete description of the velocity and displacement field during pile driving. The soil behaviour during driving and during dynamic and quasi-static testing after driving is evaluated, in order to gather information about the relation between static and dynamic soil resistance.

The third test shows the dynamic stresses in the skeleton by using a photo-elastic method. The sample consists of crushed glass, saturated with oil. The behaviour of this mixture is more or less comparable with the behaviour of saturated sand. The dynamic deviator stress is visualised by using a high speed film camera.

5.2 Pile-driving simulator

5.2.1 Introduction

The numerical simulation of the wave propagation in dry and saturated soil under a dynamically loaded pile reveals a significant difference [Hölscher, 1992]. This difference can be explained by the relatively low compressibility of the water in the soil. In saturated soil the energy radiates in a cone under the pile-toe. In dry soil the cone is significantly wider due to the importance of the shear modulus relative to the bulk modulus.

The target of the experiment described in this section is to verify the numerical results and to measure the radiation cone in dry and saturated sand under a pile during driving.

5.2.2 Experimental set-up

In this section the experimental set-up and the measurement devices are described. The turning of the pile up-side-down needs special attention.

Mechanical part of the set-up

Figure 5.1 shows the mechanical part of the experimental set-up.

A ball rolls in a pipe (1) and hits the wooden cushion (2a) on the head of the pile (3a). The pile (hard pvc) is placed in a casing (3b). Only the pile toe has contact to the soil (4). The space between the pile and the casing is covered with a membrane (4b), made from liquid latex. The pile is kept in position against the soil by four springs (2b). The soil sample is placed in a small pvc container (4a) and covered with a plastic sheet (4c). This sheet is glued to the container in order to reach an almost closed box. The sample is kept at low pressure by a vacuum pump (4d). For the wet test the sample is saturated with water by a free tube (4e). The motion of the surface during dynamic loading is measured by optical displacement transducers (5b) and monitored indirectly by a camera (5a). The velocity of the ball (1) is also measured (5c).

The precise dimensions are shown in Appendix 5A. Figure 5.2 shows a photograph of the set-up.

Measurement technique

The pile and the soil were put up-side-down. This offers the possibility of using an optical measurement method at an interface 'under' the pile toe.
In Figure 5.3.a the problem is sketched in the normal position together with the radiation areas in dry and saturated soil. The use of a number of accelerometers leads to problems

1. The angle of the radiation cone is not known accurately from the numerical analysis. In order to find the cone shape accurately a number of accelerometers is required.
2. Since the test concerns a scale test, the disturbance of the accelerometers their self and the connection wires will be large.

Figure 5.3.b shows the set-up with two optical devices mounted above the soil sample. These devices measure the soil displacement locally without influencing the soil. Only a few points of the soil surface can be considered, depending of the number of devices at hand. The entire surface can be measured by repeating the test a number of times with the position of the device slightly changed each time. This requires a high reproducibility of the test.

Another possibility is using a camera and take pictures of the surface motion or film the surface motion. Since the motion of the surface is small and takes place in a very short time interval, it is hard to obtain a good accuracy.

In this research the motion of the surface was visualised by putting a small ridge of dry sand on the surface of the sample, separated by a thin impervious membrane. Due to the wave reflection at the surface the dry sand jumps up. This is shown in Figure 5.4.

The jump-motion of this sand is easier to observe. The magnitude of this motion is determined by the initial velocity and the duration of the motion is determined by the

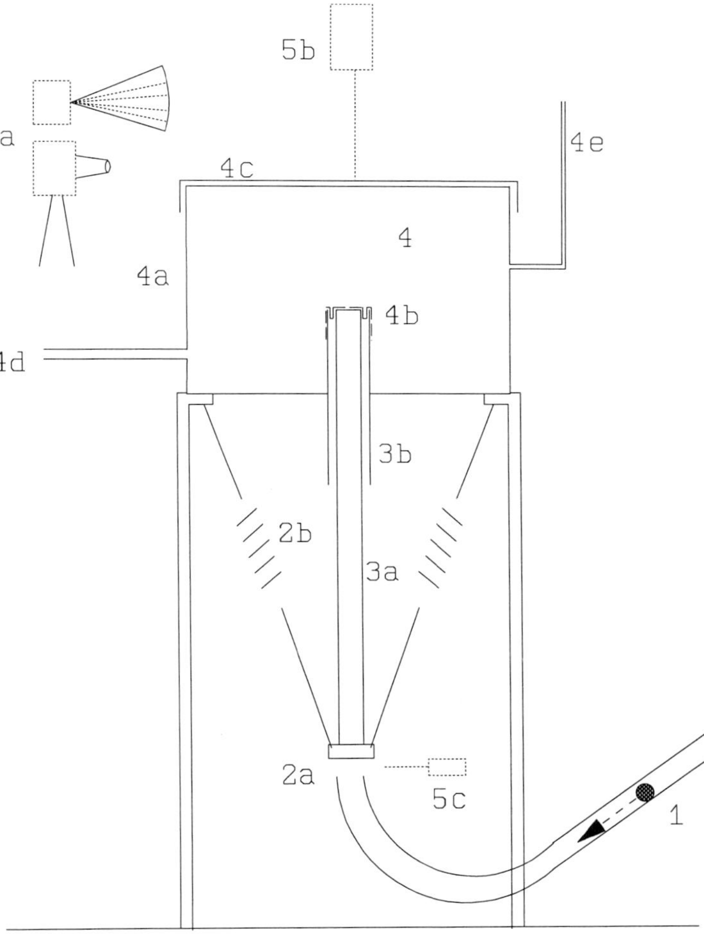

Figure 5.1 Mechanical part of set-up (explanation of numbers in text)

acceleration due to gravity (in stead of the stiffness of the soil). In this case this leads to a sixfold slower phenomenon with a twentyfold larger displacement. In Appendix 5B this is explained more in detail.

The initial velocity of the dry sand ridge is related to the intensity of the wave arriving at the soil surface along the sand ridge. In general the energy carried by a wave is proportional to the squared particle velocity. The initial velocity of the dry sand ridge is proportional to the particle velocity. The maximum height, to which the sand ridge is shot, (the jump height) is proportional to the square root of the initial velocity. Thus the local jump height of the sand ridge is proportional to the energy in the wave reflecting at that location.

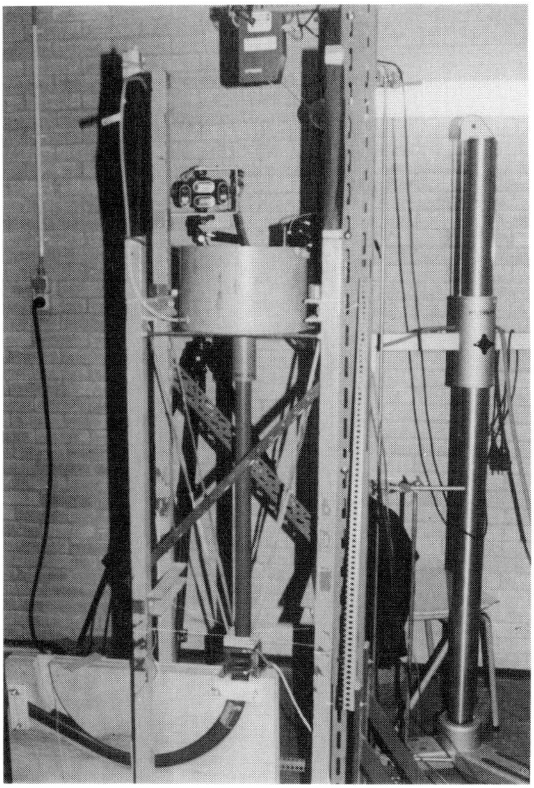

Figure 5.2 Experimental set-up

Measurement devices

Two measurement devices were used (c.f. Figure 5.1):

1. A device (5c) to measure ball speed.
2. A device (5b) to measure the motion of the soil surface (actually: the motion of the thin dry sand layer).

The device to measure ballspeed is shown in Figure 5.5. It consists of two LED's, placed at a known distance. The ballspeed is obtained by measuring the time needed for the ball to move from the lower LED to the upper LED. This method is insensitive for the possibility that the centre of gravity of the ball is not exactly in the centre of the LED's. The error due to a motion which is not exactly parallel with the line through the LED is at maximum 6 %. A typical time T_{LED} is 10.6 ms. Since L_{LED} equals 60 mm, one finds v_{ball} is 5.7 m/s.

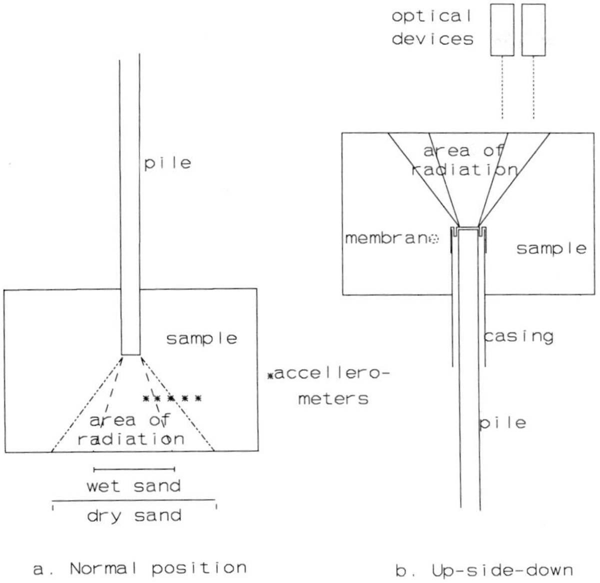

Figure 5.3 Hypothesis to be verified and the use of an optical device

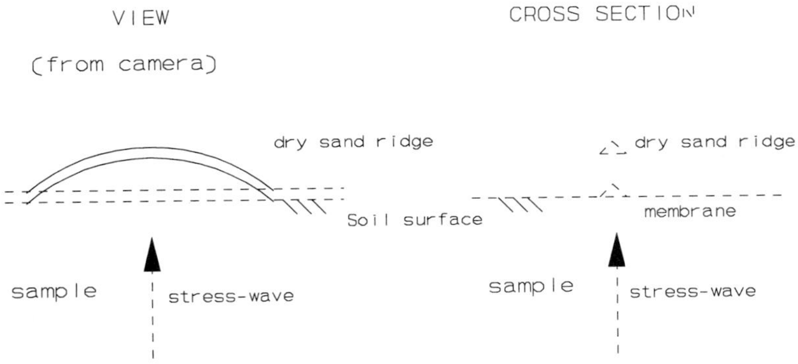

Figure 5.4 Dry sand ridge for measurement of the surface motion

The motion of the soil surface was measured directly with two optical displacement transducers and indirectly by the motion of the dry sand ridge. The direct measurement are not very accurate, due to the short time it takes place (c.f. Appendix 5B).

The motion of the dry sand ridge is activated by the reflection of the stress wave at the soil surface. In the first tests the motion of the sand ridge was photographed, but in successive tests it was recorded on high speed film, which gave better results.

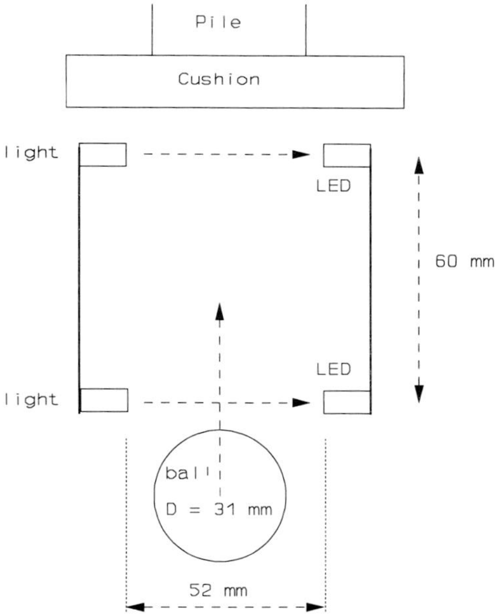

Figure 5.5 Device for measuring ballspeed

The photographs were made in a dark room with the lens opened continuously. Four flashes (Metz) were triggered by a LED at the pile head. Each flash had its own delay time (25, 35, 45 and 55 ms; the time between the flashes was 10 ms). Consequently, one picture shows the motion of the sand ridge at four successive stages. Since the duration of a flash of an ordinary flashlight is in the order of several ms, the time between two flashes cannot be decreased. Using an instamatic camera, the pictures are available within a few minutes after testing. An example of a picture is shown in Figure 5.9. The photographs show a reasonable view of the general process, but it is hard to derive the exact phenomena.

The highspeed film was made by the photographic department of Delft Hydraulics, using a 'Hyspeed' camera. The speed chosen was 2000 pictures/s. The sample was illuminated by 8 bulbs of 300 W each (total 2400 W). The film was developed and the relevant part was selected and transferred to video. The video was analyzed using the image processing system TIM [Allersma, 1991]. This postprocessing will be described afterwards.

5.2.3 Test procedure

The test was carried out for four cases:

- wet and densely packed
- dry and densely packed
- wet and loosely packed
- dry and loosely packed

Sample preparation

Eastern Scheld sand was used for the tests. The porosity of the sand can be in the range 38-48 % [Stuit, 1992].

The dry densely packed sample was prepared in thin layers (about 1 cm each). Each layer was tamped by a steel block (area 31*31 mm^2, height 125 mm). When the surface was about 3 cm above the desired level, the surface was carefully trimmed to the required height

From a measurement of the weight and the volume of one dry sample a porosity of about 39% was found. The relative density is estimated to:

$$D_r = \frac{n_{max} - n}{n_{max} - n_{min}} = \frac{0.48 - 0.39}{0.48 - 0.38} = 0.90 = 90\% \tag{5.1}$$

Thus, the sample was very dense.

The wet densely packed sample was prepared in the same way as the dry densely packed sample, but the layers were tamped submerged in water. After trimming, a membrane was installed to prevent evaporation.

The dry loosely packed sample was prepared by using a high permeable sieve. This sieve was placed at the pile toe. The box was filled with dry sand, up to a few centimetres above the final level. The sieve was pulled slowly out of the sand and the surface was carefully trimmed. This means that the sand above the level of the pile toe was loosely packed, but under the level of the pile toe the sand might be somewhat denser.

The wet loosely packed sample was prepared by slowly pluviating dry sand in the boxed, which had been filled with water. Afterwards the surface was carefully trimmed.

After removing the sand above the desired level, a thin plastic sheet was put on the sample and this sheet was glued to the container. Afterwards, the springs which kept the pile in place were connected to the cushion. At this moment the total force in the springs was about twice the deadweight of the pile.

A small ridge of dry sand (c.f. Figure 5.4) was carefully placed in the middle of the sample just in front of the measurement scale.

Test sequence

When a vacuum pump was used, a suction of about 60 to 80 kPa was applied. The sample was pumped vacuum before the test. During the test the pump had been switched of.

5.2.4 Interpretation of results

Determination of jump height from film recording

Once the film had been copied on a video tape and a time device (clock) had been added, the pictures were processed by the image processing program (TIM). The main target was to find the distribution of the maximum jump height of the dry sand ridge on top of the sample. Figure 5.6 shows a print of the video. It shows 8 stages in clockwise direction.

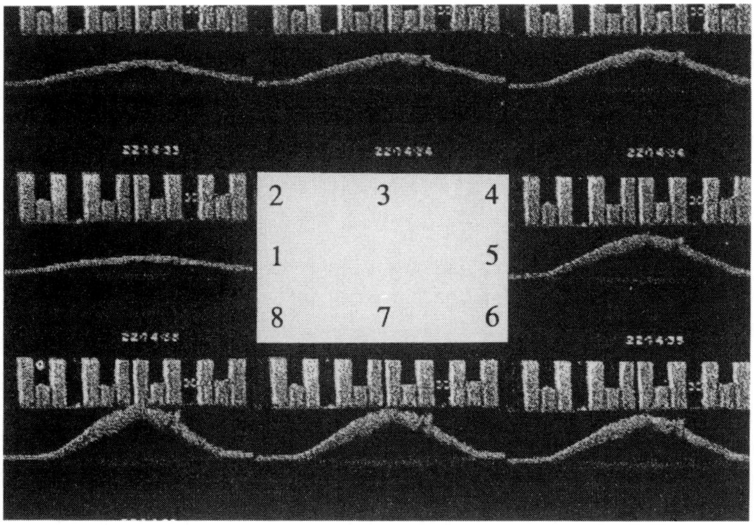

Figure 5.6 Typical video pictures

By using the image processing program, the vertical position of the upper boundary of the sand ridge was calculated as a function of time. This was done for several positions along the sand ridge.

Figure 5.7 shows the motion of the sand ridge as a function of time for several distances from the centre of the sample. These time histories contain some noise due to the vibration of the camera and the digitalization of the picture. The noise is much smaller than the measured heights. The sudden drop in height during the tests for (dry) loose sand samples is due to the digitizing process.

From many of these time histories the maximum height was obtained for several distances from the middle of the sample. These maximum heights are shown in Figure 5.8 on the vertical axis as a function of the distance (on the horizontal axis). Some graphs in Figure 5.8 show 2 lines, since two different blows were considered. The graphs in Figure 5.8 were shifted in horizontal direction and rotated a few degrees in order to obtain symmetrical graphs.

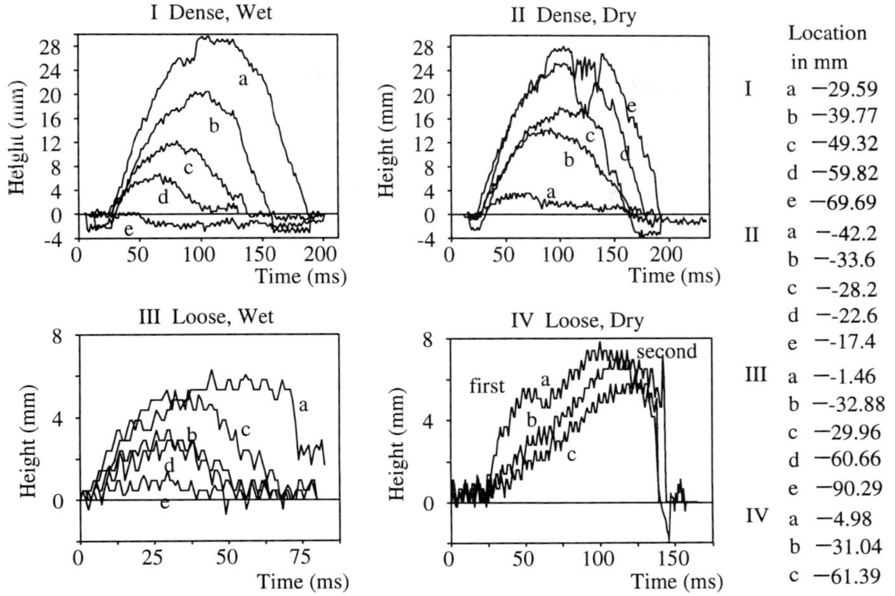

Figure 5.7 Motion of sand ridge depending on time

<u>Differences between photos and film</u>

Now, it is possible to discuss the difference between photos and film, in order to answer the question why the result could not be achieved by using photos. Figure 5.9 shows an example of the resulting pictures of a test on dry sand and one on wet sand.

Figure 5.10 shows the jump height of another test based on the highspeed film. The vertical dashed lines show the minimum time between two flashes (10 ms). The Figure shows that at different positions the maximum height is reached at different moments. In order to find the width of the zone with motion of the dry sand ridge a short time between the flashes is desired and the measurement must take place immediately after the blow, shown at the top of the Figure by the numbers 1 to 4. In order to find the height of the motion a longer time between the flashes is desired and the measurement take place longer after the blow. These requirements are hard to fulfil simultaneously. From Figure 5.10 it is derived that a time between the flashes of approximate 20 ms gives a good opportunity to see all maxima. However, with such a low frequency the small motion at the ends of the jumping part of the sand ridge cannot been seen.

A close comparison of the resulting maximum jump height shows that the pictures essentially reveal the same behaviour as the high speed film. The results based on the film give more confidence, as the measurement was performed with a much higher sample frequency.

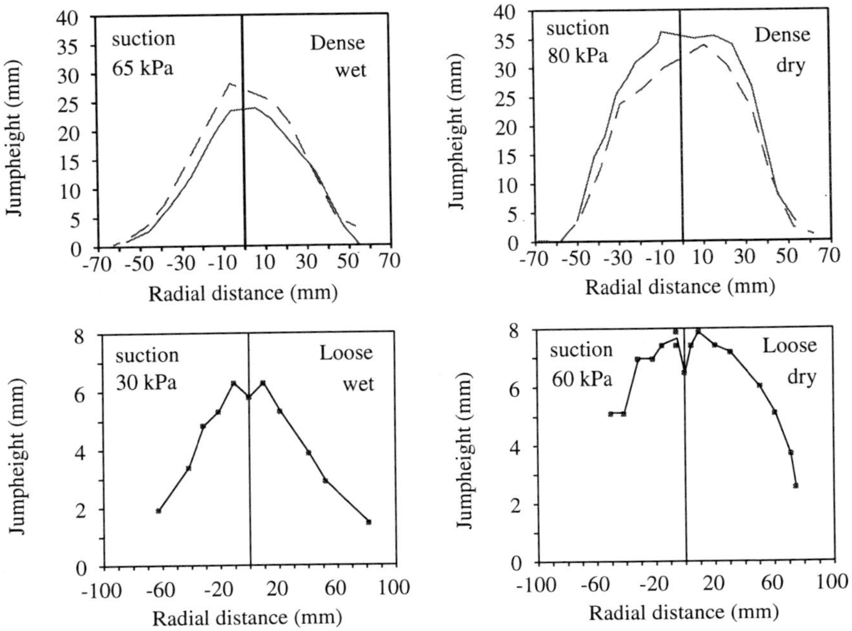

Figure 5.8 Maximum jumpheight depending on distance

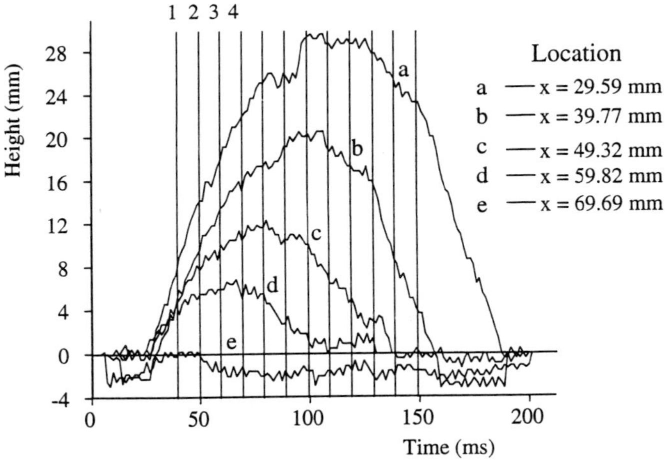

Figure 5.10 Jumpheight depending on time and possible sampling using photographs

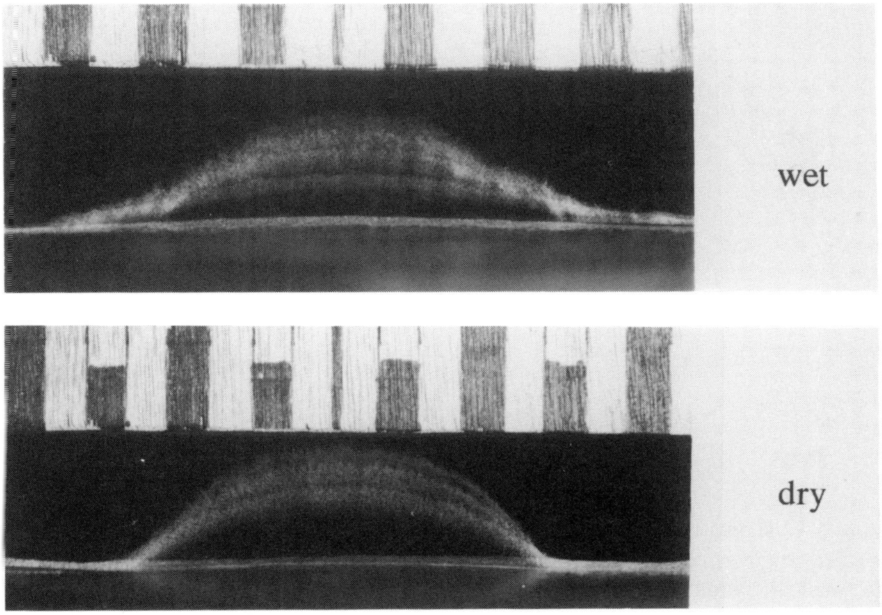

Figure 5.9 *Photograph of the jumping sand ridge lens open, flashes at 25, 35, 45 and 55 ms*

Differences between dense dry sand and dense wet sand

The area of energy radiation under the pile toe is in dense dry sand wider than in dense wet sand. This is shown by the area in which the dry sand ridge jumps up. Figure 5.11 shows the maximum jump height on radial distance for several tests with dense dry sand together with those for dense wet sand with suction and without suction. On the wet sand the region in which the soil jumps is smaller than on the dry sand.

It is mentioned that a dry test cannot be carried out without suction, because the strength of the dry sand at the surface is to low. The wet sand creates some strength.

Estimation of cone angle of radiation areas in dense sand

The cone angle of the radiation area can be estimated from the measurements. The accuracy is not high, since it is not always clear where jumping motion of the sand ridge starts exactly, due to the noise of camera vibration and digitising.

Figure 5.12 shows the jumpheight for the wet and the dry sample (two tests each) both with suction. The position of the pile toe is drawn. The maximum jump height is exaggerated by a factor 2.

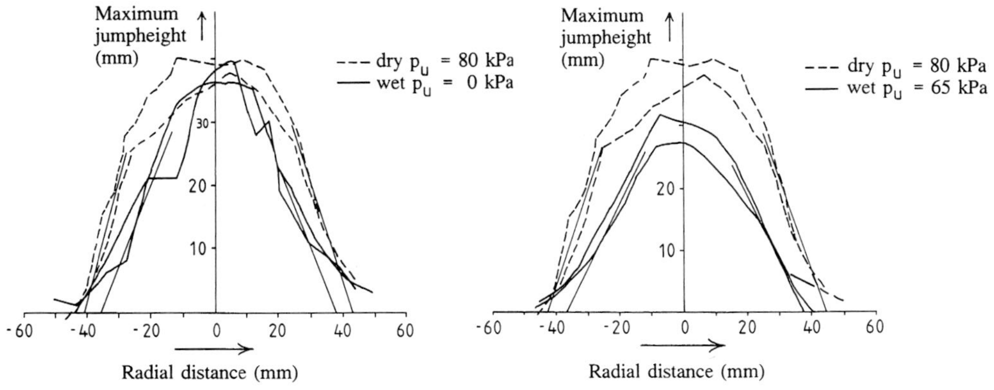

Figure 5.11 Comparison jumpheight in dense dry sample and in dense wet sample

Figure 5.12 shows that the cone angle in the dry soil is about 23° and the wet soil about 15°. This angle is determined by linear extrapolation of the curve with the steepest slope. The small deviation of this linearised curve and the measured curve outside the radiation area might be due to the radiation of shear waves, which run parallel to the soil surface. For wet soil the value coincides with the calculated value of 12° [Hölscher,1992]. For dry soil the value of the cone angle was not calculated.

<u>Differences between loose dry sand and loose wet sand</u>

The results of the test on loose dry sand and loose wet sand are hard to interpret.

In the dry sand two phenomena seem to occur. The first phenomenon relates to a stress wave, which leads to a jump of the sand ridge in a very limited area (radius approximately 20 mm, this equals the pile radius). The second phenomena causes a complete heave of the sand surface over a large area (radius approximately 100 mm). The total heave is larger than the jump height related to the first phenomenon.

Figure 5.13 shows two pictures of a test on a loose dry sample. In the upper picture the small jump can be seen. In the lower picture (75 ms later) the complete heave can be seen.

The behaviour of the loose dry sand was strongly non-linear. The final displacement of the pile toe was 35 mm, limited by the maximum admissible displacement of the membrane over the pile toe and the casing. However, a heave of about 1 mm of the surface was observed after the test. In the loose wet sample the final motion of the pile toe was only a couple of millimetres.

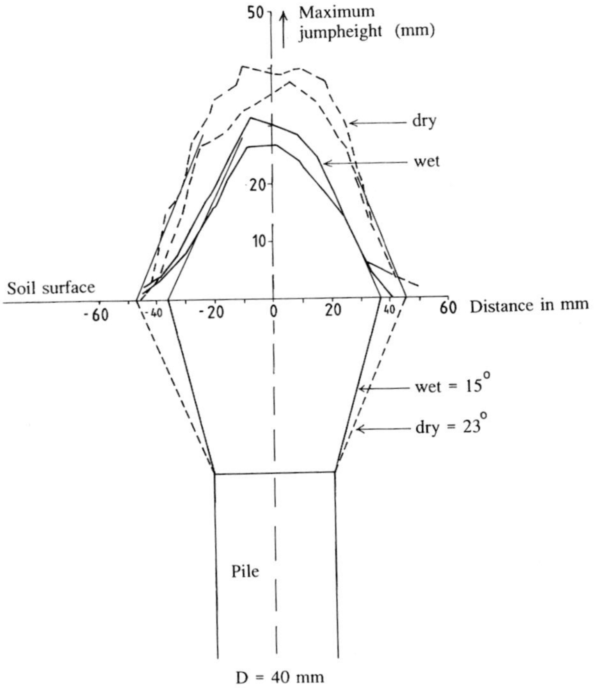

Figure 5.12 Estimation of cone angle of radiation area with suction 65 kPa

Comparison between loose wet sand and dense wet sand

Comparison of the dense wet and loose wet sample reveals two differences:

- The sand ridge on the dense wet sample jumps much higher
- The sand ridge on the loose wet sample jumps over a larger area

This means that in the loose sample less energy radiates over a larger area. Since the loose sample is expected to have a smaller shear modulus, this conclusion is against the initial theory.

5.2.5 Discussion

The reproducibility of the blow (from the ball) is good. The intensity of the blow can be derived from the ball speed measurements. The reproducibility of the sand sample condition is not studied extensively. The reproducibility of the total test itself seems fair, the test results do not change very much after a number of blows. Since the aim of the test is a qualitative prove, it is not necessary to pay more attention to the reproducibility.

For the dense sample the difference between the dry and the wet sample is as expected. Since the final displacements of the pile are small, non-linearity is expected to have small

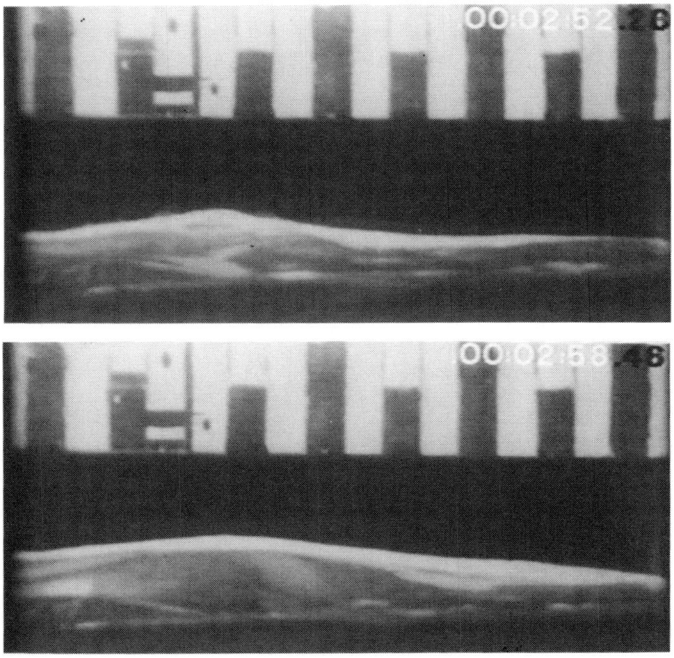

Figure 5.13 Photographs of a test with loosely packed dry sand

influences.

The tests with loose samples cannot easily be interpreted, since the soil behaviour seems to be strongly influenced by non-linearity. The influence of the strength cannot be excluded.

5.3 In-situ measurement near the toe of a dynamically loaded pile

5.3.1 Introduction

At the Fourth International Conference on the Application of Stress Waves in Piles a large number of field tests have been carried out [Geerling and Smits, 1994]. One of these tests included the measurement of pore water pressure and accelerations in the soil close to a pile with a high sample rate during pile driving, dynamic loading and quasi-static and the measurement of pore water pressure and soil pressure with a lower sample rate during static loading. The measurements with the high sample rate are presented in this section. The measurements carried out during the static loading are not studied in this thesis.

5.3.2 Description of measurements

Site conditions and pile

The soil on the test site is a soft soil, which is typical for the western part of the Netherlands. The cone resistance during a cone penetration test is shown in Figure 5.14. From the surface up to approximately 15 m depth soft soil (silty clay, silty sand and peat) is found, with a few, thin sand layers. At 15 m below surface a solid sand layer starts. This sand layer usually supports the pile foundations of buildings.

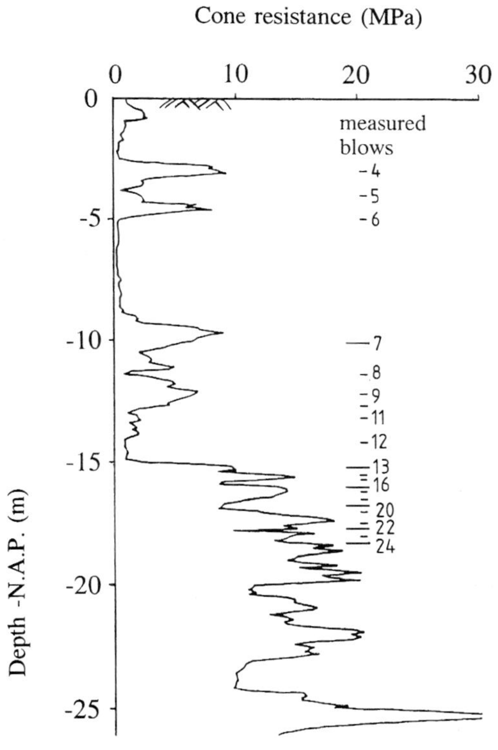

Figure 5.14 Result cone penetration test and position of toe at measurements

The test pile was a pre-fabricated reinforced concrete pile, with cross section 25*25 cm^2. The head and the toe were instrumented. The pile was driven by a IHC hammer (ICE SC-40).

Instrumentation of the soil

Three sets of each three (orthogonal) accelerometers, and four water-pressure transducers had been installed in the soil. Their location was close to the final position of the pile toe. Installation had taken place before pile driving. The location of the transducers is shown in Figure 5.15 and Table 5.1. Three water-pressure transducers had been installed around the pile, just above the final pile level. Two sets of accelerometers had been installed at approximately the same depth. The fourth water-pressure transducer and the third set of accelerometers had been installed in one group, 20 cm under the final pile level. In order to carry out measurements very close to the pile toe without disturbance of the soil close to the pile, a tube (diameter 15 cm) had been installed with an inclination of 60°. Through this tube the transducers were pushed an additional 1.3 m into the soil to reach their final position.

Table 5.1 Position of transducers

transducer	number	distance to pile (m)	depth from surface (m)	type
water-pressure	1	0.7	17.1	piezo resistance
	2	1.0	17.1	piezo resistance
	3	1.0	16.2	piezo resistance
	4	0.5	18.4	piezo resistance
accelero-meter	1	1.5	17.2	servo electric
	2	1.5	17.2	servo electric
	3	0.5	18.4	servo electric
soil-pressure	1	0.8	18.9	static
	2	1.6	18.9	static

The water-pressure transducers were piezo-resistive transducers. Calibration shows a linear behaviour up to 1000 Hz. The accelerometers were servo-electric. They are linear up to 300 Hz. Higher frequencies were filtered out by a low-pass filter.

During driving water pressure transducers 2, 3 and 4 didn't function. During the quasi-static loading the functioning of transducer 4 could be restored.

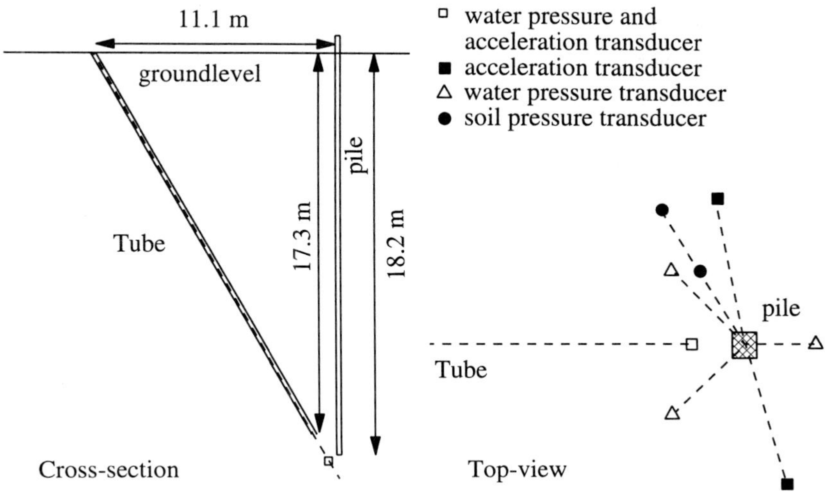

Figure 5.15 Position of transducers

Measurement program

The test pile was driven into the sand layer at 15 m. The final depth of the toe was 18.2 m below the surface: the toe penetrated 3.2 m into the sand layer. During pile driving 21 measurements (of 2 s) were carried out for several depths of the pile toe. Most of these measurements took place during the penetration into the deep sand layer. Figure 5.14 shows the position of the pile toe when the measurements were carried out.

Three days after driving a dynamic test was conducted. During three dynamic tests measurements (of 8 sec.) were carried out.

Two days later, the test pile was tested quasi-static. The test equipment used has been described by Janes, Horvath and Bermingham [1991]. During quasi-static (statnamic) testing one measurement (of 8 sec.) was carried out.

5.3.3 Results of measurements during pile driving

Pore water pressure

The water pressures are compared for several depths of the pile toe. Figure 5.16 shows the measured water pressures, together with the corresponding location of the pile toe.

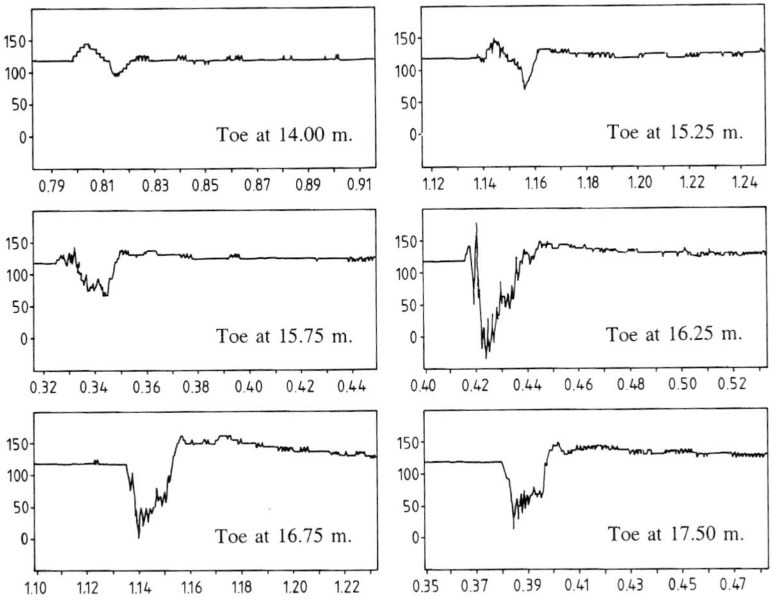

Figure 5.16 Water pressure during driving for several depths of the pile toe
Horizontal: time (s); Vertical: pore pressure (kPa)

At a distance of 1.0-1.5 m between the pile toe and the pore water transducer the character of the measured pore water pressure changes.

When the pile toe is far away, firstly compression and secondly decompression occurs (compression is positive). When the pile toe is close to the transducer, firstly decompression occurs and secondly compression occurs.

This observation shows that the compressional wave generated by the blow in the soil, propagates vertically in a limited zone under the pile toe (see Figure 5.17). The phenomenon of a limited radiation zone has been found earlier from finite elements calculations [Hölscher, 1992].

Accelerations

The measured acceleration signals were processed in order to make interpretation possible. Each accelerometer device consists of three orthogonal transducers.

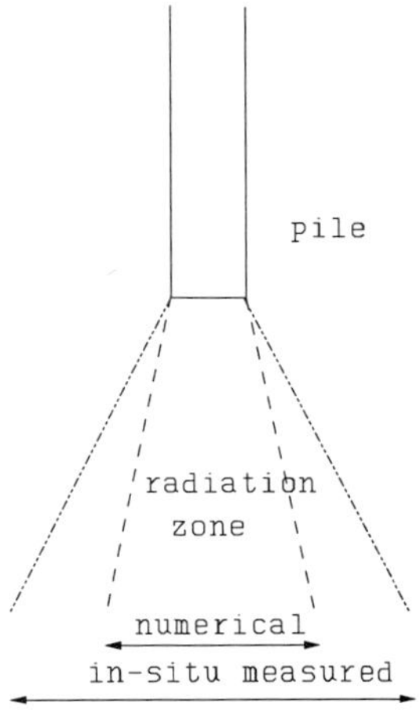

Figure 5.17 Sketch of radiation cone under the pile toe

Firstly, the measured signals were rotated in order to get signals in an cylindrical coordinate system r, z, ϕ with the origin at the pile axis. The r-axis is horizontal in the direction to the transducer, the z-axis is vertical downward and the ϕ-axis is horizontal perpendicular to the direction to the transducer. The signals of the transducers which had been installed through the tube were rotated over 30°. The signals of all transducers were rotated around the vertical axis (z-axis), so that the ϕ-component is almost zero. These rotations were supposed to be independent of the depth of the pile toe.

Secondly, the measured signals were integrated with respect to time, after subtraction of the mean value. This leads to velocity signals. These velocity signals were integrated with respect to time once more, after subtraction of the mean value of the velocity before the part of interest starts.

From the measurements synthetic velocity and displacement fields can be derived for several moments. This interpretation of the measured data is based on the assumption that the response of one transducer for blows with the pile toe at several distances from the transducer, represents the response of the soil at different distances from the pile toe for a single blow. Evaluation of measurements from successive blows then results in a comprehensive response field for one blow. The validation of this approach depends on:

- the homogeneity of the soil
- the variation of the energy per blow
- the (highest) wave velocity in the soil

In the sand layer the homogeneity is more or less justified by Figure 5.14. The variation of energy per blow is small, as can be derived from Geerling and Smits [1994]. The wave velocity is needed to synchronize the measured signals. This is needed, since the measurements in soil are not synchronised with the measurements in the pile.

Figure 5.18 shows the composed velocity field (v_r, v_z) at three moments with intervals of 15 ms. Since the velocity calculated from the signals in the two devices at 1.5 m from the pile, are almost equal, only one velocity vector is shown in Figure 5.18.

Figure 5.18a shows large particle velocity in the transducer at 0.5 m from the pile under the level of the pile toe. This velocity may be related to the compressional wave. The

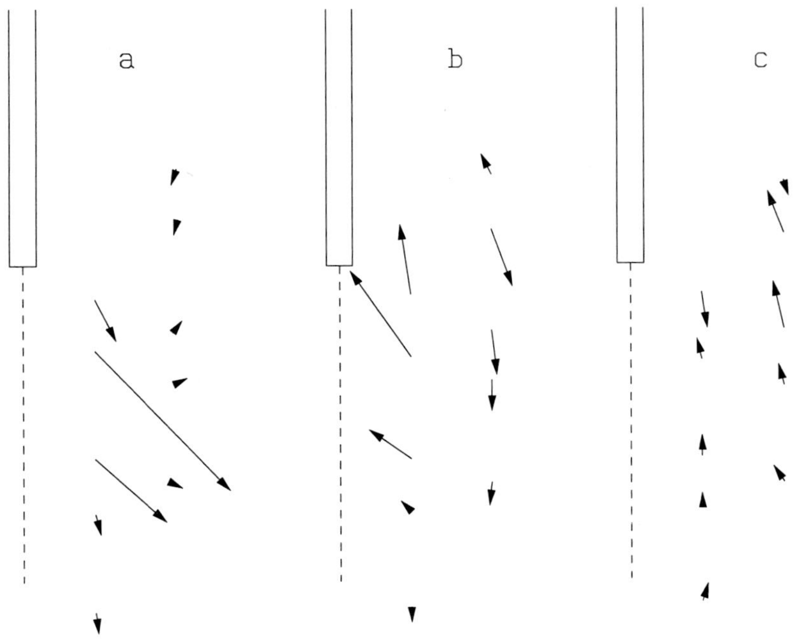

Figure 5.18 Composed velocity field in sand layer

horizontal (radial) component is quite large. This might be related to the penetration of the pile into the soil.
15 ms later, Figure 5.18b, the soil close to the pile toe moves in upward direction. This reflects a significant elastic rebound. The vertical downward particle velocity (in Figure 5.18a) propagates in horizontal direction. This can be seen from the large downward particle velocity at the transducer(s) at 1.5 m from the pile.

30 ms later, Figure 5.18c, the soil near the pile toe moves down, with a very low particle velocity. However, the upward particle velocity of the toe (in Figure 5.18b) propagates in horizontal direction resulting in an upward velocity at the transducer(s) at 1.5 m from the pile.

Figure 5.19 shows the composed displacement field at six moments at 10, 20, 30 40, 50 and 60 ms. This Figure shows similar phenomena as Figure 5.18. The final displacement cannot be found due to noise in the signals.

5.3.4 Dynamic testing

After driving the pile was tested dynamically. For this test the same hammer as for pile driving was used. The hammer was dropped on the pile, which resulted in a force of 15.6 kN in the pile head.

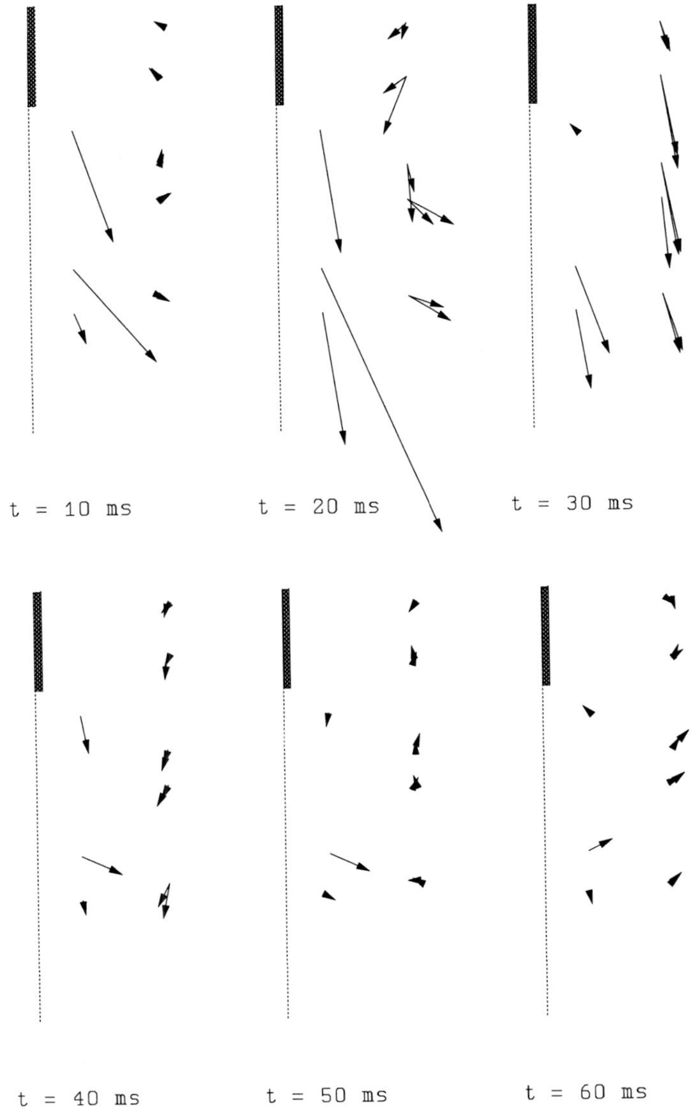

Figure 5.19 Composed displacement field in sand layer

The signals measured during the dynamic testing can be compared with the last measured signals during driving, since the pile penetrated only a few centimetre after this measurement.

Figure 5.20 shows the velocity signal which were measured by the transducers nearby the pile for the last driving blow and the first dynamic test. The characteristic response in both situations is similar, particularly the arrival of the wave (downward-upward motion). The attenuation is different, maybe due to different plastic deformation. The difference in vertical and horizontal velocity is considerable. This difference cannot be explained from the difference in impact force of the hammer: the energy level during dynamic testing was a little bit smaller during driving. A difference in stiffness of strength of the soil around the pile tip and a difference in shaft resistance might be the cause.

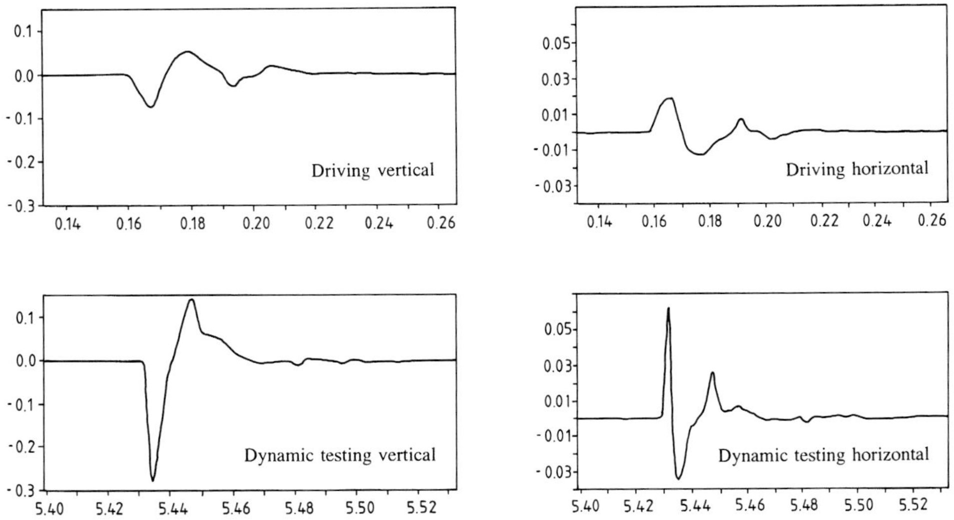

Figure 5.20 Comparison velocity during driving and dynamic testing
Vertical: Velocity (m/s); Horizontal: Time (s)

For the interpretation one has to bear in mind that the acceleration transducer is 20 cm under the pile level of the pile toe and approximately 50 cm beside the toe. According to the findings in the previous Section, all transducers are outside the radiation area which exists under the pile toe during the last blow of pile driving and during dynamic testing.

5.3.5 Quasi-static testing

A couple of days later, the pile was tested quasi-statically by a so called statnamic test.

Figure 5.21 shows the measured water pressure for the last blow of pile driving, the dynamic test and the statnamic test.

From Figure 5.21, it is observed that the duration of the quasi-static loading was longer than the duration of the dynamic loading. After transformation to the frequency domain the main frequency from dynamic loading is 30 Hz and for quasi-static loading it is 8 Hz.

Further, it is observed that the time dependency is almost equal: firstly, the pore-water pressure decreases and, secondly, it increases. For pile driving this behaviour is explained by the fact that the compression wave under the pile toe radiates in a limited region under

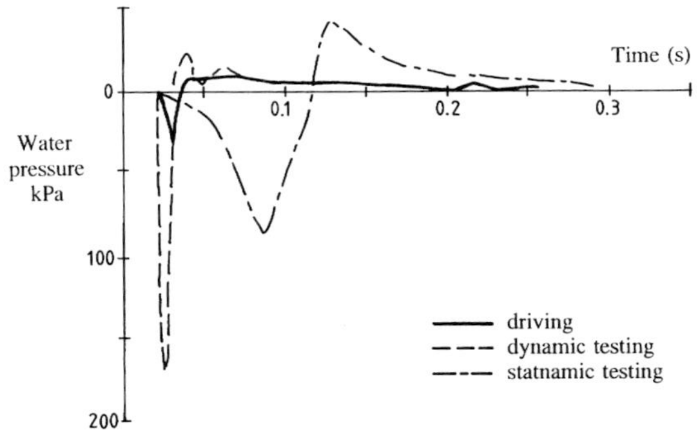

Figure 5.21 *Comparison of water pressure during pile driving, dynamic testing and statnamic testing*

the pile toe (see previous Section). Above the pile level of the pile toe (where the pore-water transducer is installed), extension wave propagates into the soil. Extension waves lead to a decrease of pore-water pressure first, followed by an increase due to the elastic rebound.

During the quasi-static loading large pore water pressure occurred. In order to give a good estimation of the importance of the pore water pressure, the force on the pile head during the statnamic test should be compared with the force during the dynamic test. Unfortunately, the force during the statnamic test is not available.

The consolidation time of the soil around the pile toe is between 100 and 200 ms, estimated from the measurements of pore-water pressure during driving. This time is relatively long in relation to the loading time from the statnamic loading. Therefore, static pore water pressure generation might be expected. This implies that for the determination of the static bearing capacity, the results of a quasi-static test must be adjusted for the influence of the dynamic pore water pressure.

Figure 5.22 shows the (almost vertical) velocity which was measured at 0.5 m and 1.5 m from the pile centre. From this Figure, it can be seen that the maximum velocity is reached at different moments: at 0.5 m at t = 1.179 s and at 1.5 m at t = 1.190 s. The time difference is small, but it relates to a wave velocity of $c = \Delta x/\Delta t = 1.0/0.012 = 80$ m/s. This wave velocity is reasonable for the shear wave velocity in the sand layer. The soil is loaded by a propagating wave and cannot be considered as statically loaded.

5.3.6 Conclusions

Based on these measurements it is concluded:

- Comprehensive measurement of the dynamic soil motion near the pile toe during driving is possible.
- The results are coherent in a physical sense. They provide a possibility to test soil models which are commonly used in pile dynamics.

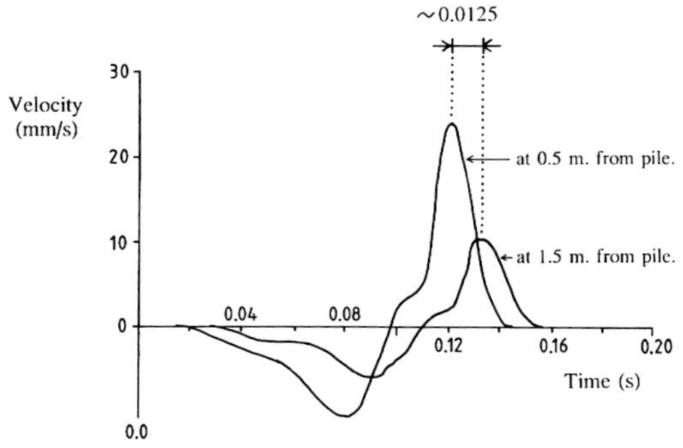

Figure 5.22 Vertical velocity during quasi-static loading

- The theory of the radiation cone under the pile toe (based on finite element calculations) is confirmed. However, the measured top-angle of this cone is bigger than the one estimated from the finite element calculations.
- During pile driving and dynamic loading the measured accelerations in the soil above the pile level of the pile toe have the same character, but the attenuation differs.
- During the quasi-static loading significant pore water pressures are generated.

5.4 Photo-elastic experiment

5.4.1 Introduction

The third test is focused on the visualisation of the stresses in the skeleton. During dynamic loading of saturated soil, the fluid stress is generally spoken much higher than the solid stress. This is a complication for direct measurement of the solid stress. The usage of a photo-elastic granular material filled with a suitable pore fluid, makes it possible to visualize the solid stress.

For static loading the method used in this research has been developed by Allersma [1987]. The granular material consists of crushed glass. The pores between the crushed glass are filled with Risella oil, which has the same refraction index as the glass.

Little photo-elastic research on stress wave propagation in the skeleton of granular material is available in literature. Trense [1970], Van der Graaf [1964], Rossmanith and Shukla [1982] and Shukla et al. [1987, 1988, 1991] have studied the propagation of shockwaves in media consisting of circular photo-elastic disks. Trense and Van der Graaf have also included the influence of pore water between the disks.

The main difference between those studies and the study which is described in this section, is the fact that, in this study, the solid is an random packing of irregular shaped particles and that the particles are much smaller than the zone of interest and the thickness of the sample. Allersma [1987] has shown that the behaviour of an assembly of crushed glass in a triaxial cell resembles strongly the behaviour of sand. Since the particles are much smaller than the zone of interest, the behaviour of one single particle cannot be distinguished. Only the behaviour of the assembly is observed.

5.4.2 Relation between light-intensity and deviatoric stress

The experiment is based on the photo-elastic properties of glass: Changes of the stress in the glass lead to changes of optical properties. The changes of optical properties can be measured by measuring the light which is transmitted through the sample. Thus, the photo-elastic particles are used as stress transducers.

The relation between the difference in mean stress and the light intensity is derived in Appendix 5C [see also Allersma, 1987]. The result is

$$B = \frac{1}{2} \sqrt{2}\, A \sqrt{1 - \cos\delta} \tag{5.2}$$

with B amplitude of the light observed by the camera
A amplitude of the polarised light after the first polaroid filter
δ the retardation angle in the sample

The retardation angle is related to the mean stresses in the sample (σ_1 and σ_2) by

$$\delta = M\,(\sigma_1 - \sigma_2) \tag{5.3}$$

with M the photo-elastic constant.

For small values of δ the relation between B and δ (Equation (5.2)) is almost linear. For higher values of δ a periodic relation is found. At the stress level which is expected in the dynamical test, this periodic behaviour does not occur.

5.4.3 Experimental set-up

Figure 5.23 shows a sketch of the experimental set-up. Figure 5.24 and Figure 5.25 show pictures of the experimental set-up.

The sample consisted of crushed glass saturated with Risella oil. The sample was placed between two glass plates with a thickness of 20 mm each at a distance of 20 mm. The sample was approximately 40 cm wide and 40 cm high.

The sample was prepared by filling the box with oil. Than, the crushed glass was pluviated carefully in the oil. Thus, a loosely packed sample was obtained. Afterwards, the air was removed by pumping the oil slowly through the sample in upward direction. The pump had been switched of before a test.

Before testing, an almost constant surface load had been applied. This statical surface loading was introduced by pushing the three blocks on the sample by a bolt. A soft spring (stiffness is 37.8 N/mm) was between the block and the bolt, in order to keep the surface

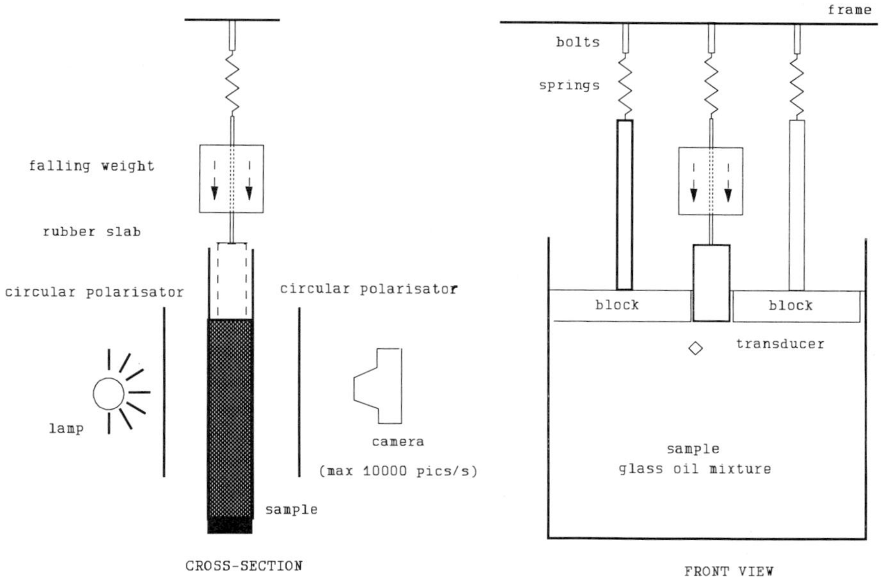

Figure 5.23 Set-up for photo-elastic experiment

Figure 5.24 Experimental set-up for photo-elastic research

Figure 5.25 Sample and surface loading on sample

loading constant during a test. The statical surface load on the loading block was introduced through a thin bar under the spring. The statical surface load was estimated from the actual length of the springs.

For a dynamical test, the dynamic load was generated by a steel block (40*40*80 mm) which felt on another steel block (18*40*80 mm). A rubber slab was between these blocks. The falling block is called here the hammer, the lower block is called the anvil.

For a statical test, the statical load was introduced by a block (18*40*80 mm) which was pushed on the sample by a bolt. A load cell had been placed between the bolt and the block.

At both sides of the sample two circular polarizer were placed. These circular polarizer existed of a polaroid filter and a quarter lambda plate. By using two circular polarizer, the transmitted light is independent of the direction of the main stresses (see Appendix 5C).

During the tests the transparency of the sample was measured, by registration of the light intensity behind the sample. This was registered by using a laser and a photo-diode, which was connected to an oscilloscope, or by using an intensive light source and a (high speed film) camera. This light source existed of 6 halogen bulbs of 1250 W each.

During the dynamic tests all 6 halogen bulbs were used and the speed of the high speed camera was 4000 pic/sec. During the statical test only two halogen bulbs were used and the light intensity was registered by using an instant camera.

The dynamical motion of the anvil was measured with an accelerometer, which had been placed in a hole in the anvil. The statical motion of the block (anvil) was measured with a marking gauge.

5.4.4 Results of measurements

Static loading

During two static loading tests the displacement of the anvil and the force on the anvil were measured. In these tests different static surface loads on the two surface blocks were applied (see Table 5.2).

Table 5.2 Static pre-load on surface for static tests

	left block area = 18*165 mm^2		right block area = 18*180 mm^2	
	force N	stress N/mm^2	force N	stress N/mm^2
test 1	230	0.077	250	0.077
test 2	300	0.101	360	0.111
picture	200	0.066	260	0.079

The resulting load-displacement curves are shown in Figure 5.26. It shows two curves for each test. After an applied displacement, the force increased directly, but during a short period afterwards, it decreased somewhat. The upper line shows the force which was reached instantaneously, the lower line gives the final force, which was reached after some time. The change of displacement during this process is not shown.From Figure 5.26 it can be seen that the failure load was almost reached. The failure load is about 2400 N. The initial stiffness is about 500 to 700 N/mm.

The unloading behaviour turned out to be very stiff, a displacement of only 0.3 mm sufficed to unload the sample.

During another static test a photograph was made. The static force on the anvil was 450 N. This photograph is shown in Figure 5.27. The surface load is shown in Table 5.2.

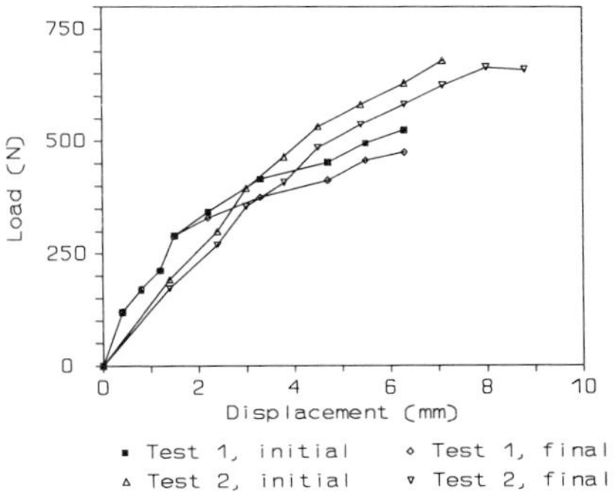

Figure 5.26 Load displacement curves static load tests

Figure 5.27 Photo-elastic effect during static loading

Dynamic loading

For several tests the light intensity during dynamic loading was measured with a laser and a photodiode and the motion of the loaded block was registered with the accelerometer. Figure 5.28 shows the measured signals for one test. The upper two parts are the measured light intensity and the measured acceleration of the anvil. The light intensity is measured about 1.5 cm under the edge of the anvil. The third curve in the figure shows the velocity of the anvil, based on the integration of the accelerometer with respect to time.
The light intensity increases in about 1 ms very rapidly, and decreases much slower during about 4 ms. The similarity between the measured light intensity in the sample and the velocity of the block is great.

The celerity of the waves can be estimated by measuring the time which the wave needs to reach a point further away. Based on measurements at 110 mm and 175 mm under the

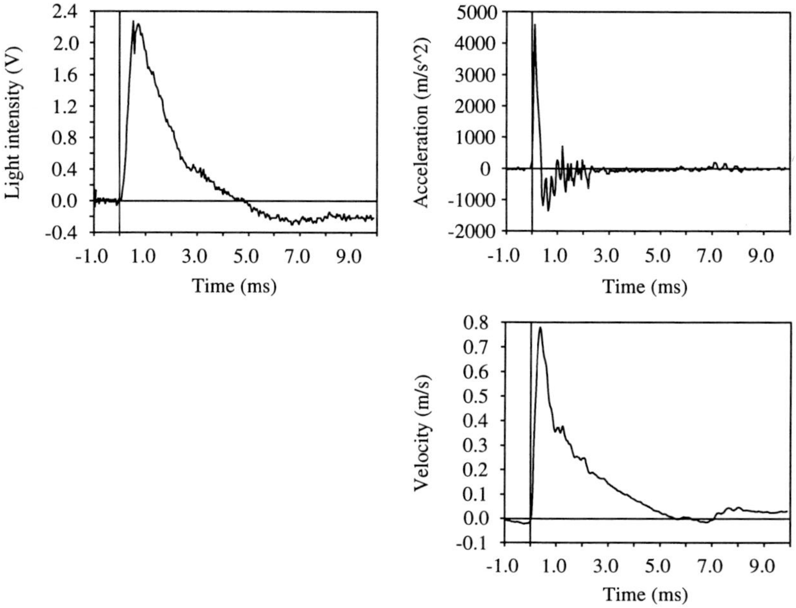

Figure 5.28 Point measurement at dynamical loading

block, the celerity of the wave is about 335 m/s.

Two dynamical tests are filmed with the high speed camera at 4000 pics/s. The static surface (pre-)loading on the two surface blocks and the load block are collected in Table 5.3. For both tests the falling height of the hammer is 7 cm. The final set of the three blocks is also shown in Table 5.3.

During the first test the anvil moved into the sample for almost 1 mm. This seems reasonable. During the second test the anvil moved up. This seems very unrealistic and points at an error during the measuring of the length of the springs.

Figure 5.29 and Figure 5.30 show the results of the filmed tests. The pictures are shown each 0.5 ms. The region, in which the light occurs, can be clearly seen. The sudden increase in light intensity, followed by a slower decrease of intensity, can be seen.

The almost stationary position of the region with high light intensity is remarkable. On first sight it seems that the deviator stress is not governed by wave propagation. However, the point measurements revealed clearly wave propagation. Therefore, it is expected that the light-intensity outside the region with high light intensity is below the threshold value of the film.

Table 5.3 Surface pre-load and final set from dynamical tests

			left	anvil	right
test 1	pre-stress	N/mm^2	0.11	0.15	0.10
	final set	mm	0.6	0.8	0.3
test 2	pre-stress	N/mm^2	0.11	0.14	0.09
	final set	mm	-0.2	-0.1(?)	0.1

5.4.5 Discussion on the influence of the bending flexibility of glass plates

The measured wave velocity (335 m/s) is lower than expected from the material properties (1580 m/s, see Appendix 5D). The reason is probably the (bending) flexibility of the glass plates. Due to this flexibility the sample is not loaded in pure plane strain. This influence is investigated in this section.

In order to simplify the investigation, the glass-oil mixture is modelled as a single phase material. The kinematic equation is substituted into the constitutive equation which yields

$$\sigma_{ij} = \lambda\, u_{k,k}\, \delta_{ij} + \mu\, (u_{i,j} + u_{j,i}) \tag{5.4}$$

with $\lambda = K_s + 2/3\ G_s$ and $\mu = G_s$ the Lamé elastic constants.

The influence of the lateral stiffness of the plate on the sample is expressed by

$$u_{3,3} = -\frac{1}{\gamma}\, \sigma_{33} \tag{5.5}$$

with x_3 perpendicular to the glass plate; γ zero means plane stress, γ large means plane strain. Substitution of Equation (5.5) into (5.4) gives for the stress σ_{11}

$$\begin{aligned} \sigma_{11} = {} & [\lambda\,(\frac{\gamma + 2\mu}{\gamma + \lambda + 2\mu}) + 2\mu\,]\, u_{1,1} \; + \\ & \lambda\,(\frac{\gamma + 2\mu}{\gamma + \lambda + 2\mu})\, u_{2,2} \end{aligned} \tag{5.6}$$

and a similar equation for σ_{22}. This equation shows that only the Lamé constant λ is influenced by the lateral stiffness of the glass plates.

In order to find the order of γ which relates to the measured decrease of wave velocity from 1580 m/s to 335 m/s two approximations are made: the shear modulus μ is much smaller than γ and the Lamé constant λ (of the mixture) is mainly determined by the compressibility of the fluid (see Appendix 5D). Then, the compressional wave velocity can be approximated as a function of γ by

$$c_p(\gamma) = \sqrt{\frac{\lambda \dfrac{\gamma + 2\mu}{\lambda + \gamma + 2\mu} - \dfrac{2}{3}\mu}{\rho}} \approx \sqrt{\frac{\lambda \dfrac{\gamma}{\gamma + \lambda}}{\rho}} \tag{5.7}$$

The value of γ can be estimated from

$$\gamma \approx \frac{c_p(\gamma)^2}{c_p(0)^2 - c_p(\gamma)^2}\,\lambda = 0.047\,\lambda \tag{5.8}$$

The value of λ is about 3800 kPa and the value of μ about 18 kPa (see chapter 6). This gives for γ a value of 178 kPa. The approximation μ much smaller than γ is thus reasonable.

The value of the stiffness γ can be estimated from the bending stiffness of the glass plate. Since the glass plate is loaded by the sample in its centre, it is considered as a beam with a simple support at both ends, which is loaded by a concentrated force. The stiffness of this beam is defined by

$$k = \frac{F}{\Delta u} = \frac{4Et^3}{L^2} \tag{5.9}$$

with	k	stiffness (N/m)
	F	(concentrated) force on the glass plate (N)
	Δu	the displacement of the glass plate (m)
	E	Youngs modulus of glass ($50*10^9$ N/m^2)
	t	the thickness of the glass plate (0.02 m)
	L	the size of the glass plate (0.4 m)

From this stiffness, the mass of the plate and the measured rate of loading, it is derived that the duration of loading is much longer than the eigenvibration time of the plate. Therefore, dynamic effects are disregarded in this approach.

For the stress in the sample perpendicular on the glassplates one can write

$$\sigma_{33} = -\frac{F}{a^2} \tag{5.10}$$

with a the size of the loaded area. For the strain of the sample perpendicular on the glassplates one can write

$$u_{3,3} = 2\frac{\Delta u}{B} \tag{5.11}$$

with B the thickness of the sample (0.02 m).

These two equations can be introduced in Equation (5.5). Together with Equation (5.9), this leads to

$$\sigma_{33} = -\frac{2Et^3B}{a^2L^2}u_{3,3} = -\gamma\, u_{3,3}$$
$$\Rightarrow \gamma = \frac{2Et^3B}{a^2L^2} \qquad (5.12)$$

Using a = 0.05 m, one finds $\gamma = 40*10^6$ N/m^2. This value is a factor 4.5 to smaller than the value estimated above.

5.4.6 Conclusion

The dynamical stress in an assembly of crushed glass can be measured by using the photo-elastic method. Accurate measurements are possible by using a laser and a photo-diode. A more general insight is possible by using high speed film.

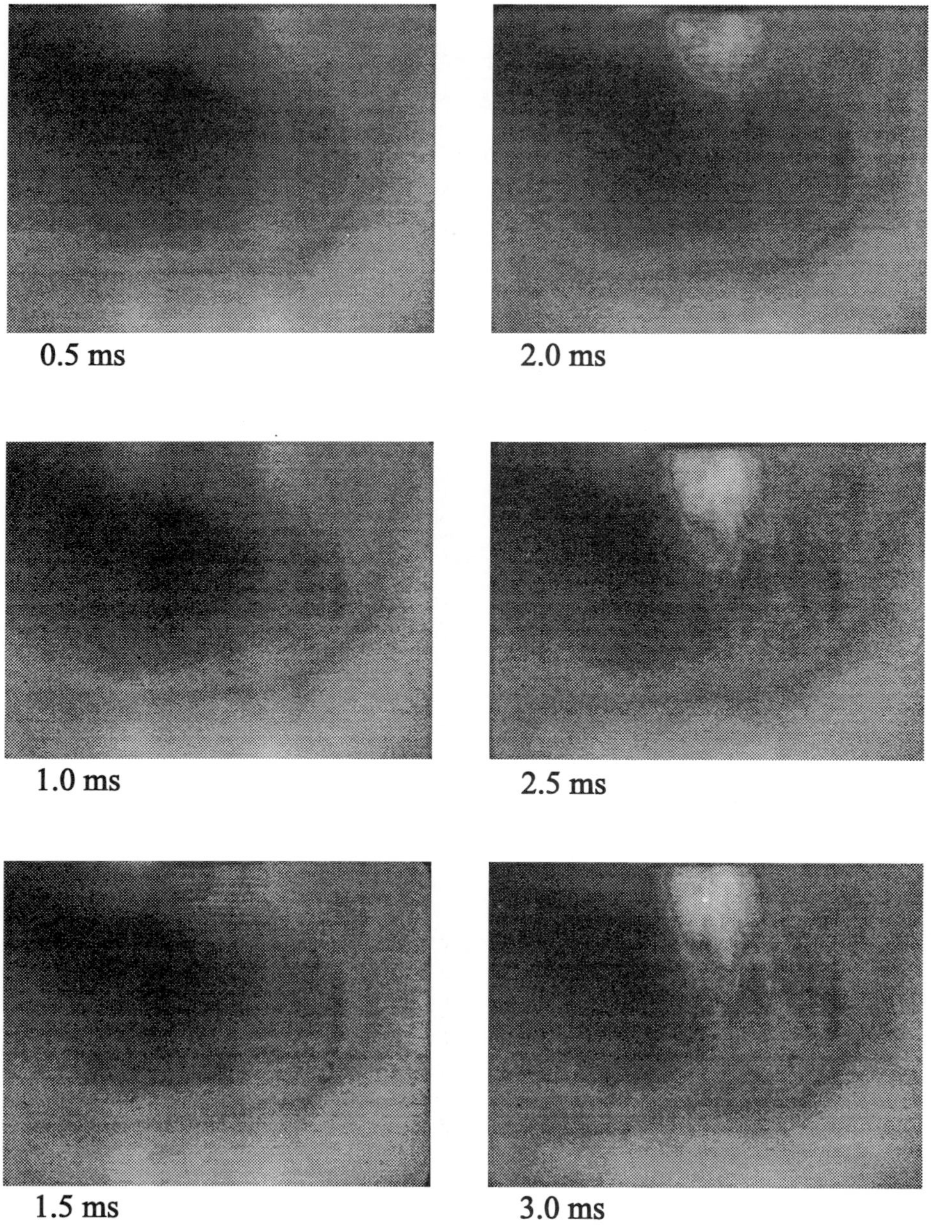

Figure 5.29 Film pictures of the first test

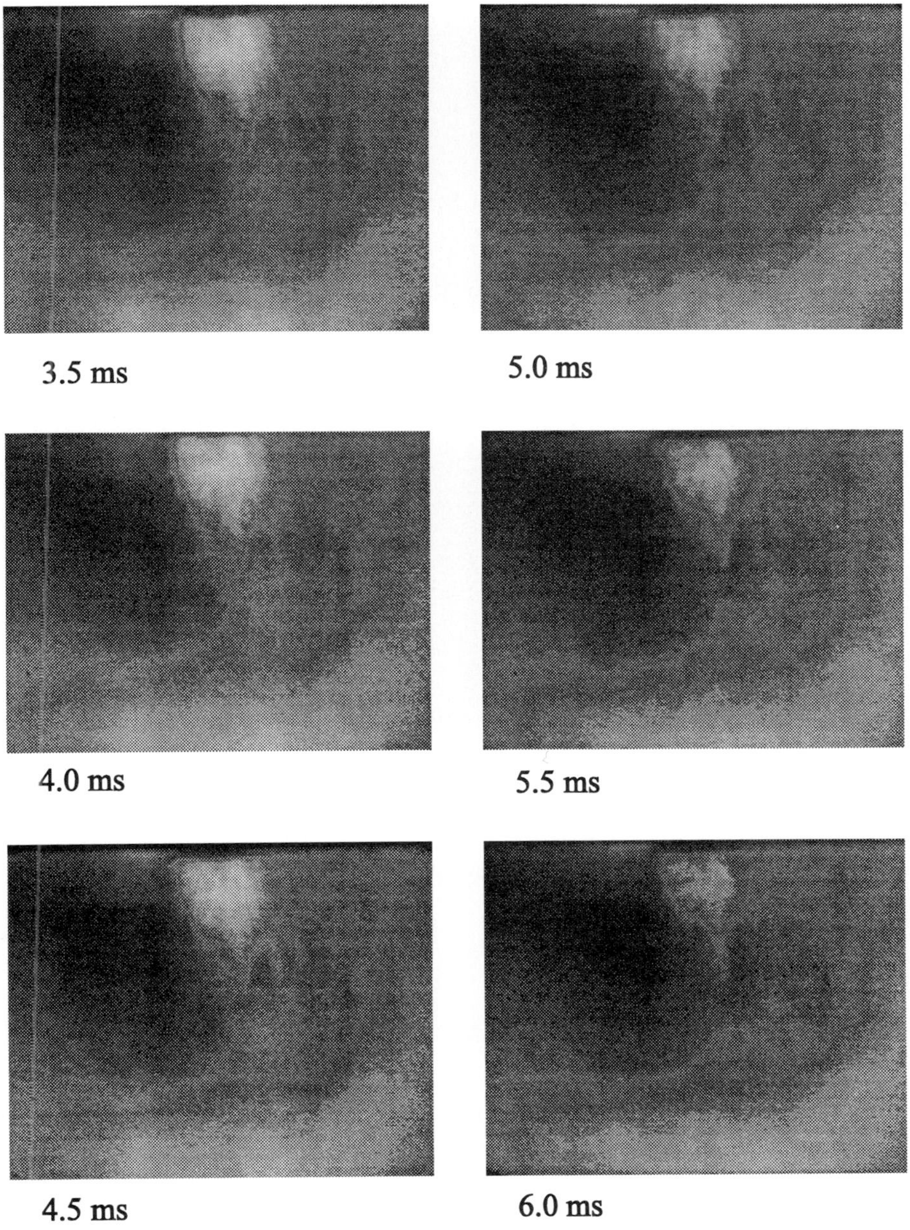

Figure 5.29 Film picture of the first test (continued)

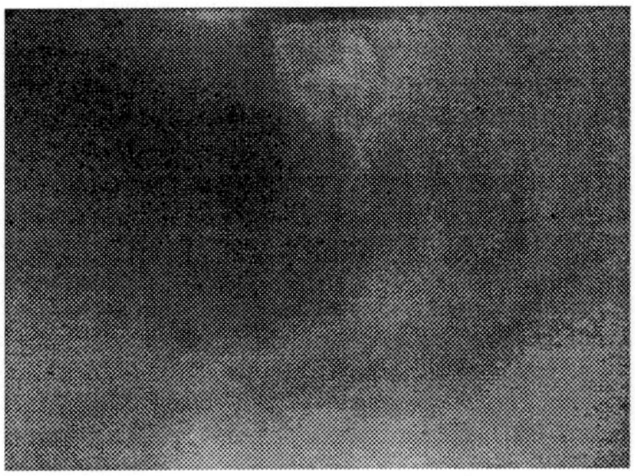

6.5 ms

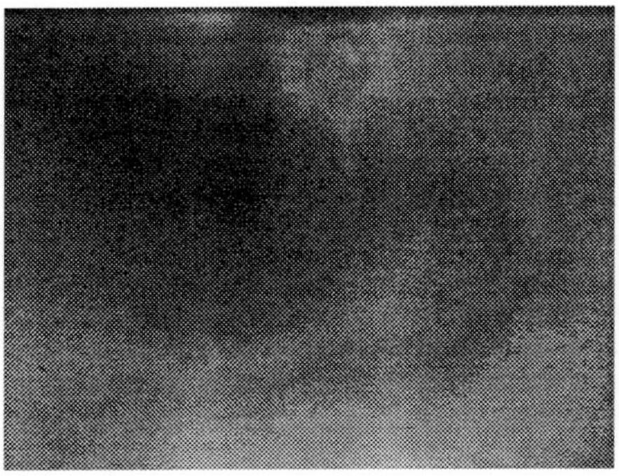

7.0 ms

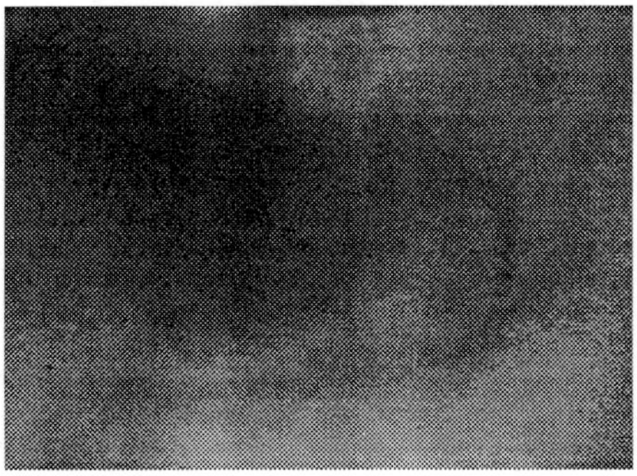

7.5 ms

8.0 ms

8.5 ms

Figure 5.29 Film picture of the first test (continued)

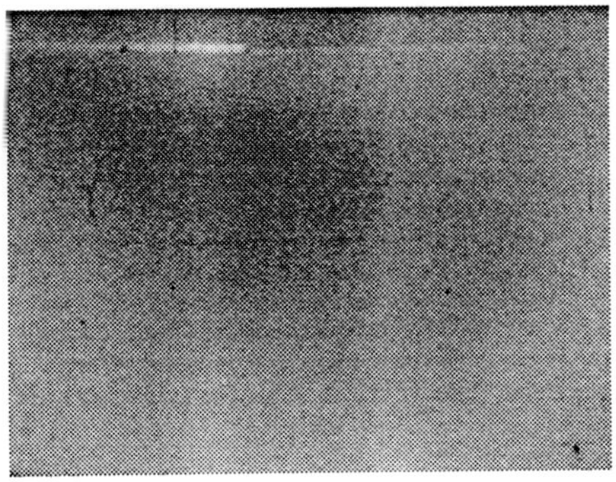

0.5 ms

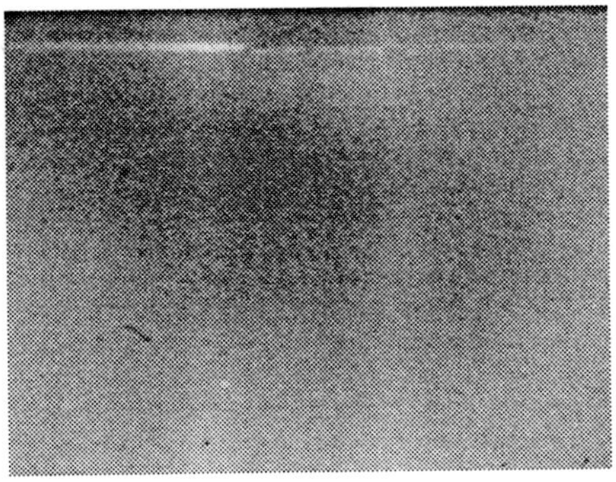

1.0 ms

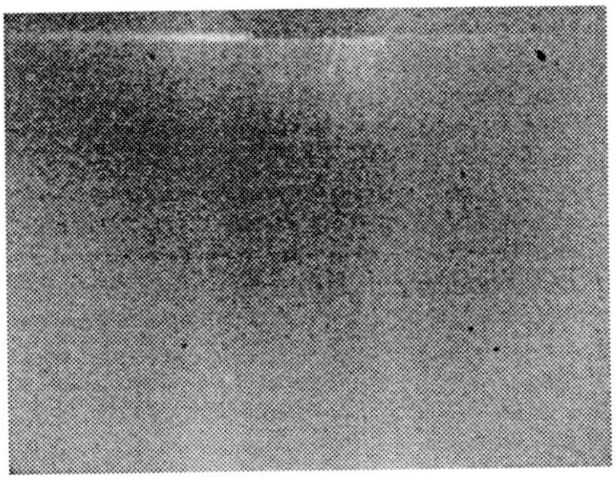

1.5 ms

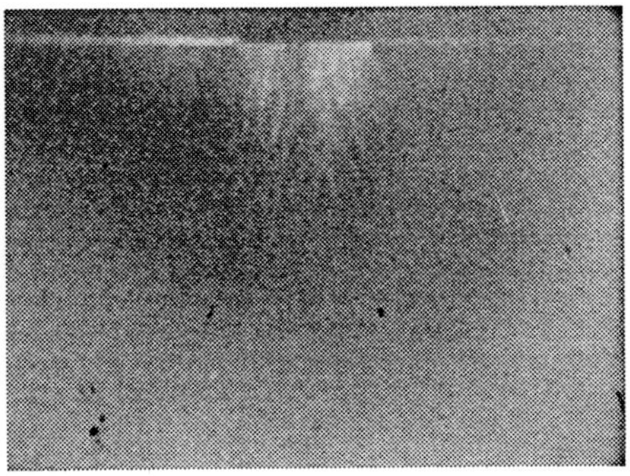

2.0 ms

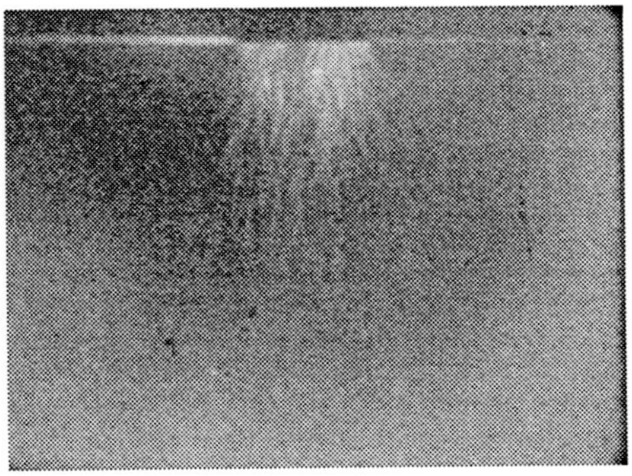

2.5 ms

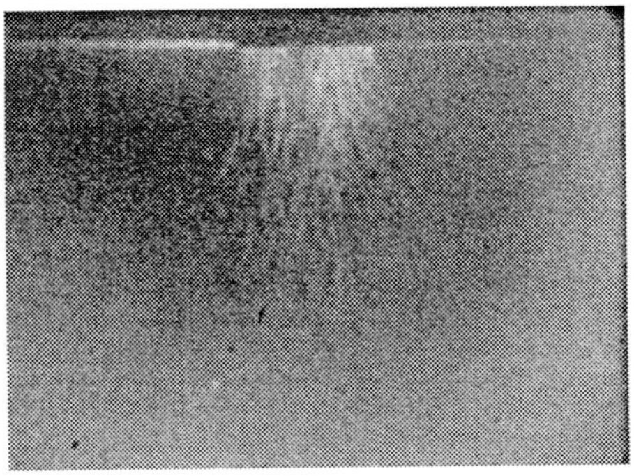

3.0 ms

Figure 5.30 Film pictures of the second test

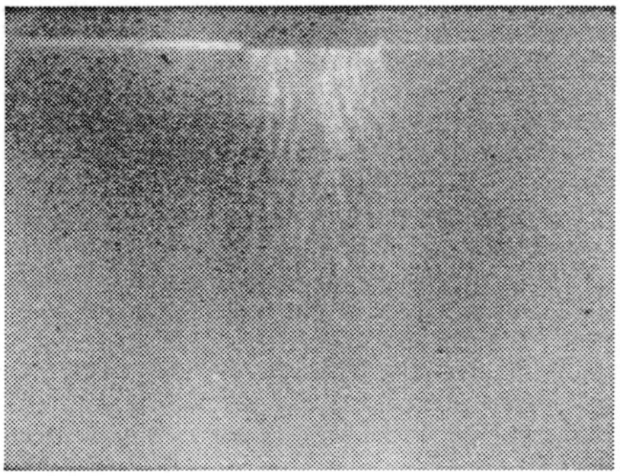

3.5 ms

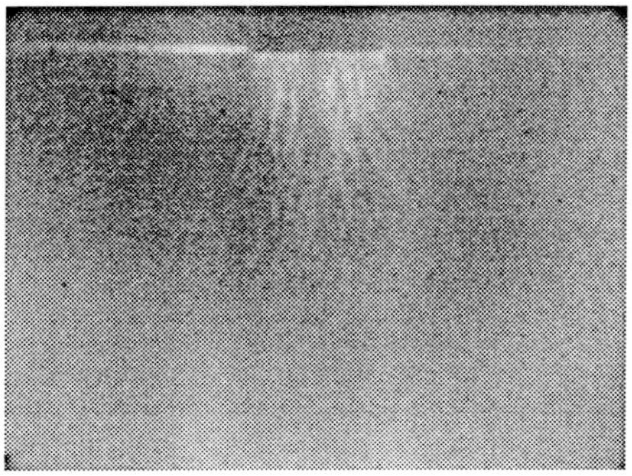

4.0 ms

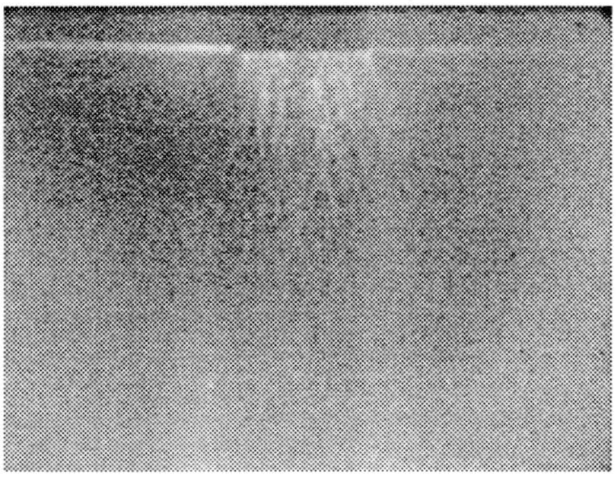

4.5 ms

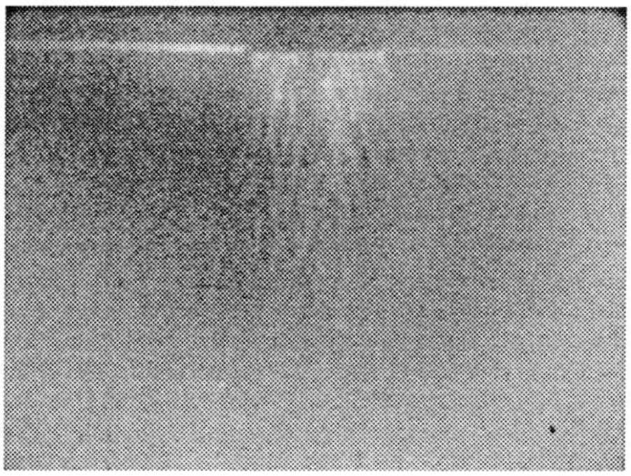

5.0 ms

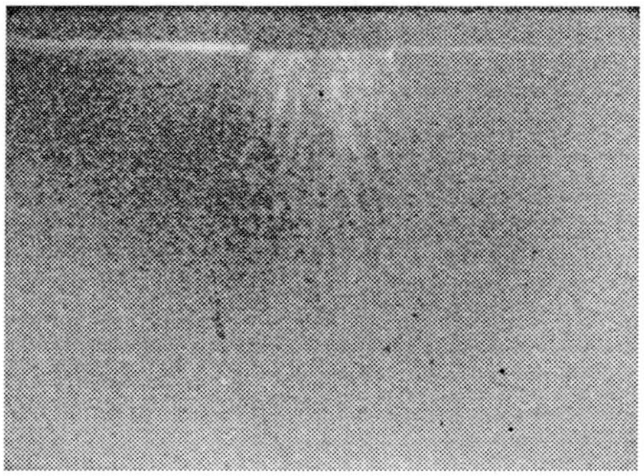

5.5 ms

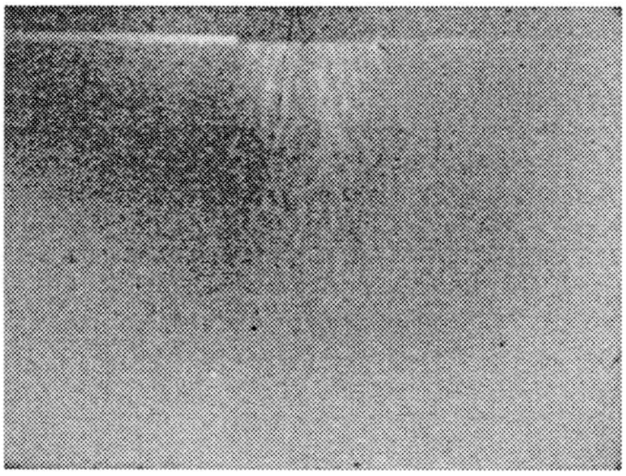

6.0 ms

Figure 5.30 Film pictures of the second test (continued)

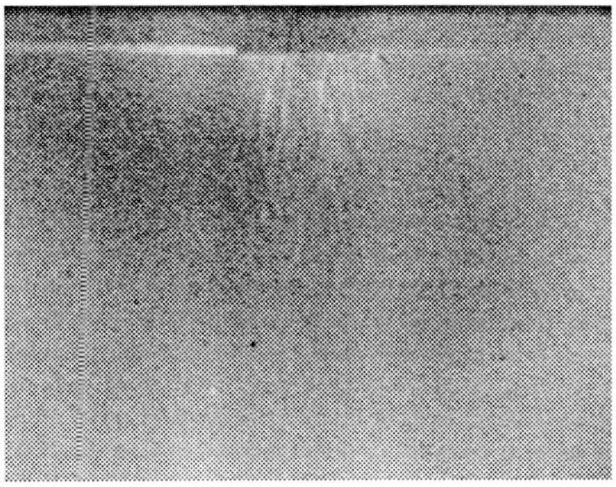

6.5 ms

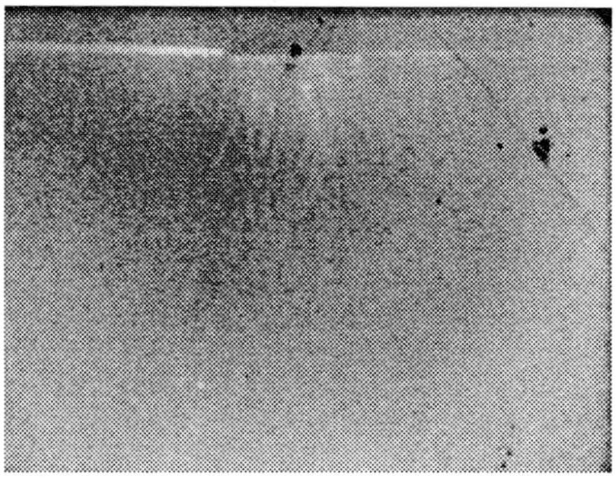

8.0 ms

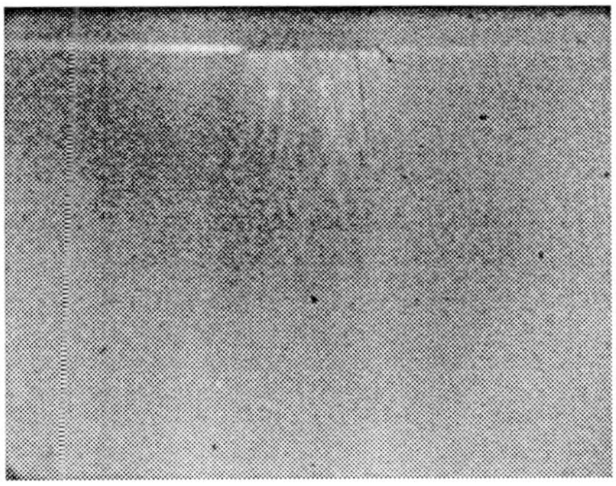

7.0 ms

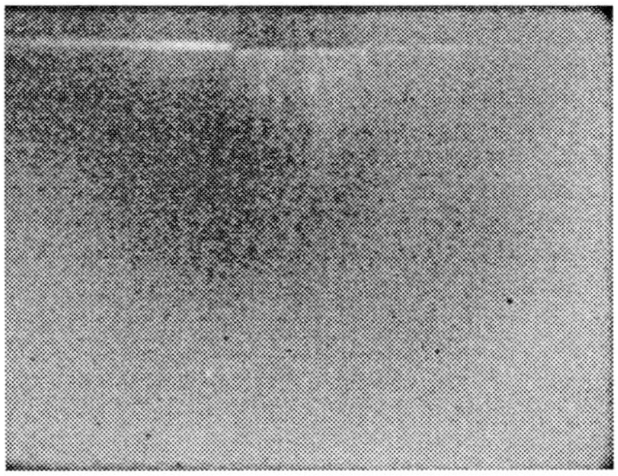

8.5 ms

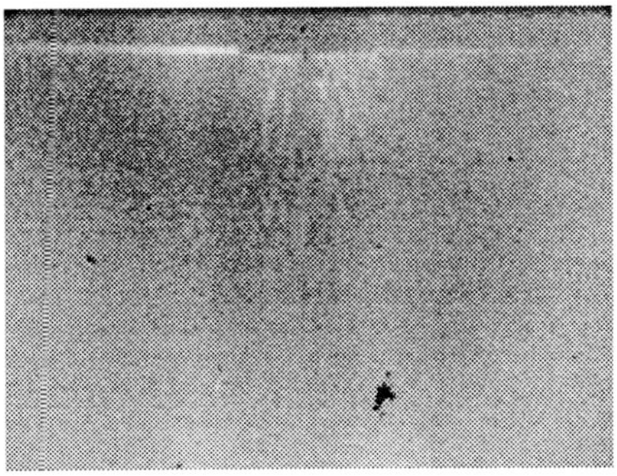

7.5 ms

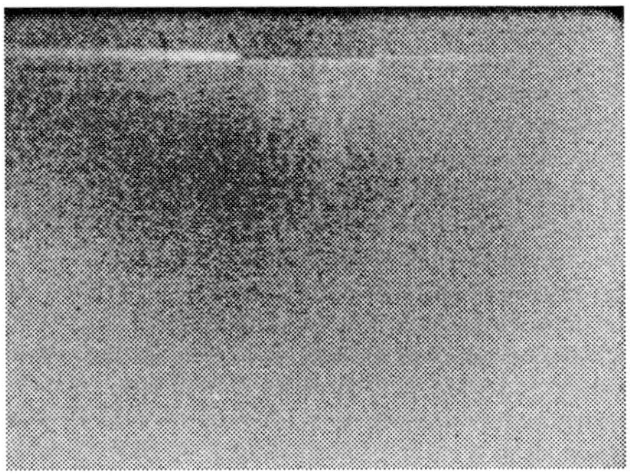

9.0 ms

Figure 5.30 Film pictures of the second test (continued)

6 NUMERICAL SIMULATION OF DYNAMICAL SOIL BEHAVIOUR

6.1 Introduction

In this chapter the finite element program TITAN is applied to the following problems:

- the pile driving simulator
- a breakwater, which is loaded by a wave blow

6.2 Pile driving simulator

6.2.1 Introduction

In this section the numerical simulation of the pile driving simulator is discussed. The test set-up is described in section 5.2, together with the experimental results.

This section focuses on two questions:
- the differences between the dry and the wet sample
- the differences between the loose and the dense sample

6.2.2 Numerical model

The problem is considered as axially symmetrical.

The material properties for the elastic simulations are shown in Table 6.1. The differences between the loose and the dense sand are estimated by using some general formulae for the dependency of permeability and stiffness on porosity.

The pile head is loaded by a force, which is triangular in time. The duration of the increasing part of the loading is estimated on 200 μs, the duration of the decreasing part of the loading on 600 μs. The maximum stress is 2.1 kN.

The boundary conditions for the sand sample are too complex to be described exactly. The behaviour depends on the stiffness of the container wall relative to the sand in the sample and the friction between the plastic container wall and the sample (stuck or moving). The stiffness of the sample depends on the test under consideration (dry or wet sand) and the kind of wave (shear or compressional wave). For simplicity the sand at the pile toe is supposed to be rigidly connected to the pile, the sand along the casing is supposed to move freely along the casing and the sand along the bottom and the wall of the container is supposed move freely along the container. The membrane on top of the sample is modelled as a free flow boundary. The other boundaries of the sample are modelled as no-flow boundaries. Figure 6.1 shows an overview of the boundary conditions.

Table 6.1 Elastic properties for simulation of the pile driving simulator

	name		unit	value	source
K_f	bulk modulus water		MPa	900	estimated
ρ_f	volumetric mass water		kg/m^3	1000	
ρ_a	apparent volumetric mass		kg/m^3	00	
ρ_s	volumetric mass sand		kg/m^3	2500	estimated
n	porosity	loose dense	-	0.45 0.35	measured
k	hydraulic permeability	loose dense	m/s	0.024 0.012	estimated
K_s	bulk modulus sand	loose dense	MPa	25 50	estimated
G_s	shear modulus sand	loose dense	MPa	10 20	estimated
K_p	bulk modulus plastic pile		MPa	834	polytechnisch zakboekje
G_p	shear modulus plastic pile		MPa	385	polytechnisch zakboekje
ρ_p	volumetric mass plastic pile		kg/m^3	1000	polytechnisch zakboekje

6.2.3 Linear simulations

During the experiments the (jump) motion of the sand ridge was measured. This motion was induced by the motion of the surface of the sample. Therefore, the velocity of the surface is studied in this section. The jump of the sand ridge starts when the velocity reaches a maximum (in time) (see Appendix 5B).

<u>Dense sand</u>

Figure 6.2 shows the calculated velocity field for the dense saturated sample at several moments. The arrival of the wave at the pile toe and the propagation into the sample can be seen.

Figure 6.3 shows the vertical velocity of the surface of the dense sample at several moments, just after arrival of the stress wave at the pile toe. Two cases are considered: saturated sand and dry sand both densely packed. From Figure 6.3 the following conclusions are drawn (on the result of first arrival at time = 1.4 ms):

- The velocity of the surface of a dry dense sample is higher than the velocity of the surface of a wet sample
- The region with a high maximum velocity of the surface of a dry sample is wider than the region with a high velocity of a wet sample.

Table 6.2 compares the measured maximum jump height (in the centre of the sample) and width with the computed height and width for the dense samples. Since the velocity is measured by an indirect method, only the maximum velocity can be calculated from the experiments.

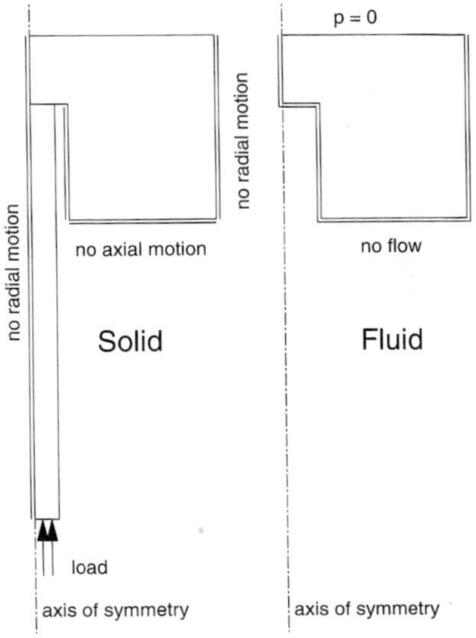

Figure 6.1 Boundary conditions

The calculated results show a tendency, which is in agreement with the experimental results. The calculated width and height are both somewhat higher than the measured width and height. This might be due to the fact that in the calculation no energy dissipates due to non-linearity. The calculated height is proportional with the estimated blow at the pile head. However, this blow is strongly dependant on the stiffness of the cushion (c.f. Appendix 6A). The stiffness of the cushion is strongly non-linear. For the target of the experiment and the calculation the differences in jump height are not relevant. The calculated width is expected to decrease due to non-linearity in both cases.

Table 6.2 Comparison experimental and computational results for dense sand

	saturated sand		dry sand	
	measured	calculated	measured	calculated
width (mm)	35	55	45	85
height (mm)	30	40	40	63

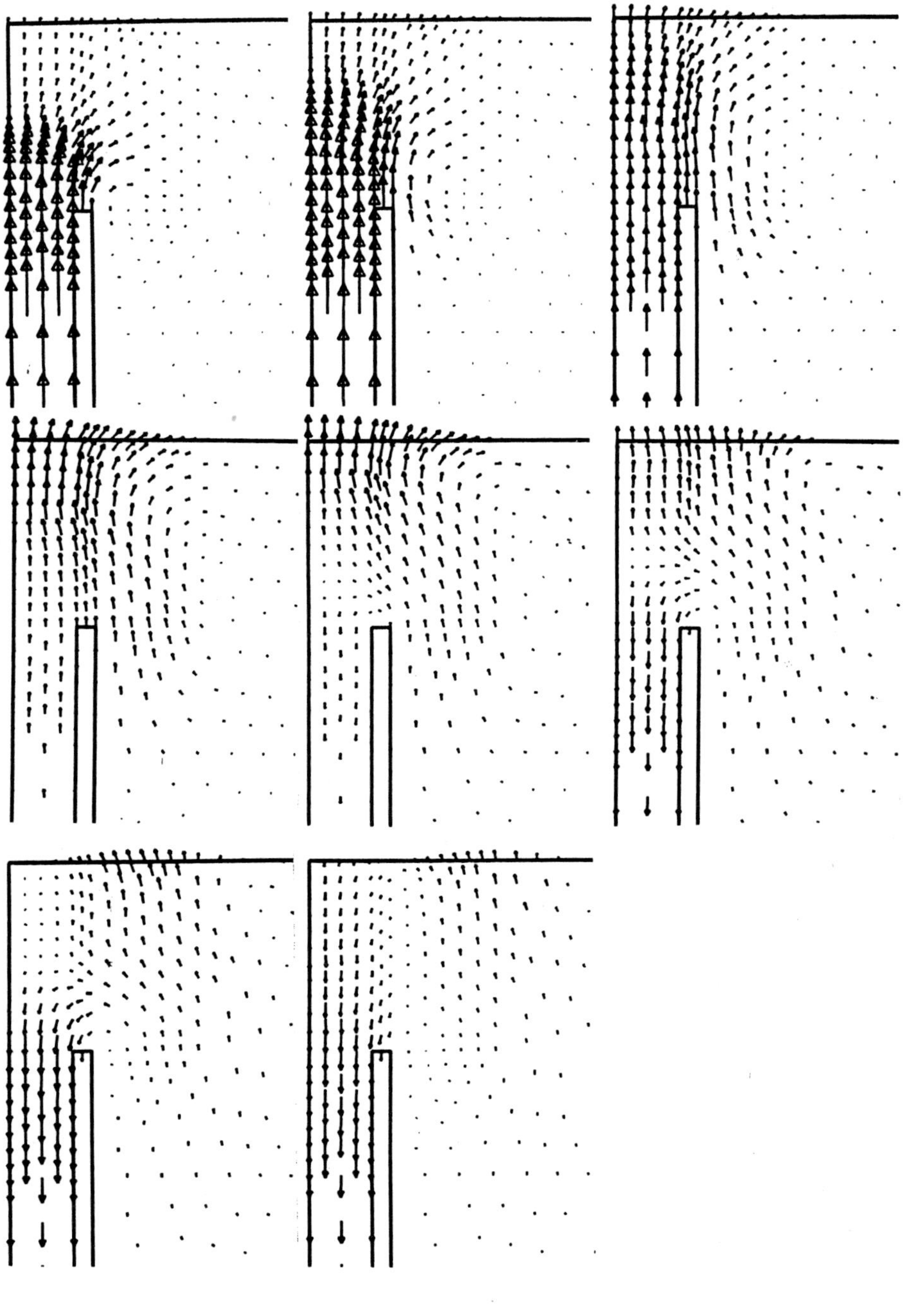

Figure 6.2 Velocity in a sample of dense saturated sand

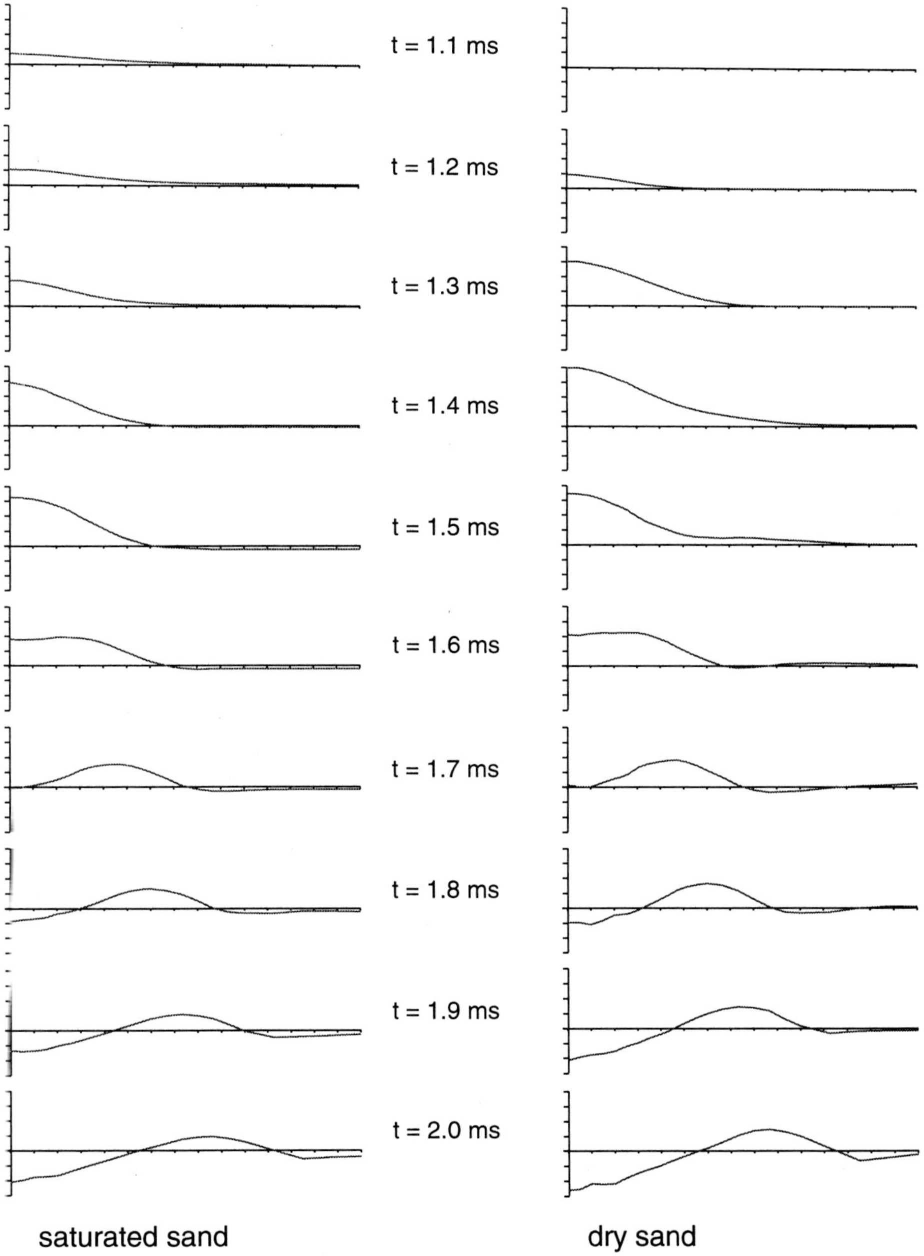

Figure 6.3 *Vertical velocity of the surface for linear material behaviour*
1 thick is 10 mm (horizontal) or 0.3 m/s (vertical)

Loose sand

Figure 6.4 shows the vertical velocity of the surface of the loose sample at several moments, just after arrival of the stress wave at the pile toe. Again, two cases are considered: saturated sand and dry sand. The results for the dense sample (already presented in Figure 6.3) are given for reference by a dotted line. Comparison shows that the results for loose sand and dense sand are more or less similar, but in the dense sample everything goes faster, due to the higher wave velocity in the dense sample.

In the experiments, a larger difference between the dense and the loose sample was found. This might be due to the fact that the stiffness of the loose sample is estimated too high and to the fact that non-linearity is neglected. During the tests the loose sample showed strong non-linear behaviour.

6.3 Simulation of the wave attack on a rubble mound breakwater

6.3.1 Introduction

During a storm, a rubble mound breakwater is loaded by two types of loading. The first one is the more or less regular upward and downward motion of the watertable at the seaside slope. The second one is the breaking of waves on the seaside slope. The breaking of waves on the slope is a dynamical loading. It is considered in this section.

A practical question for the design of breakwaters is overall geotechnical stability of the breakwater during a storm. In general, most attention is payed to the regular part of the loading. The regular wave leads to wave set-up in the breakwater and a quasi-static flow in the core and the armour of the breakwater. The simulation in this section, shows that the breaking waves generates dynamical motions of the solid, which penetrate into the entire breakwater.

6.3.2 Physical model

The breakwater considered here, is shown in Figure 6.5. The breakwater is 13 m high. The bottom of the see is horizontal. The breakwater is build up from two materials, a finer granular material in the core and a coarser material in the armour. The thickness of the armour is three meter. The side slope has an inclination of 1:3, the harbour side slope has a slope of 2:3. The width of the crest of the breakwater is 7 m.

The still water level is 7 m. Due to the regular wave motion an internal set-up of 2 m occurs at the moment of the wave blow. This internal set-up generates a declining water table in the core of the breakwater (see Figure 6.5).

For the subsoil under the breakwater, two possibilities are considered: a very stiff rock or a soft sandbed.

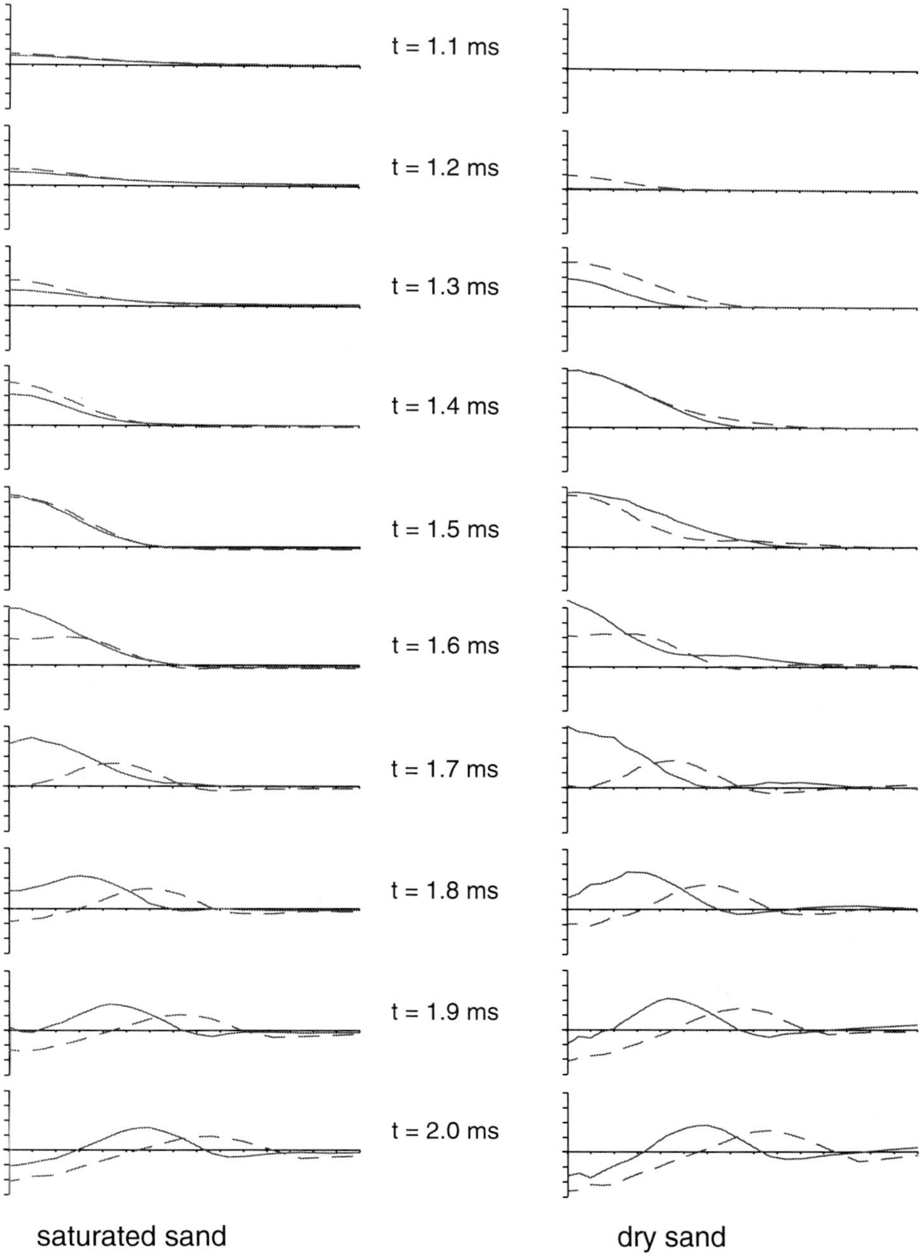

Figure 6.4 *Comparison surface velocity of simulation for saturated and dry sand full line: loose sand, dotted line: dense sand (1 thick = 10 mm ; 0.3 m/s)*

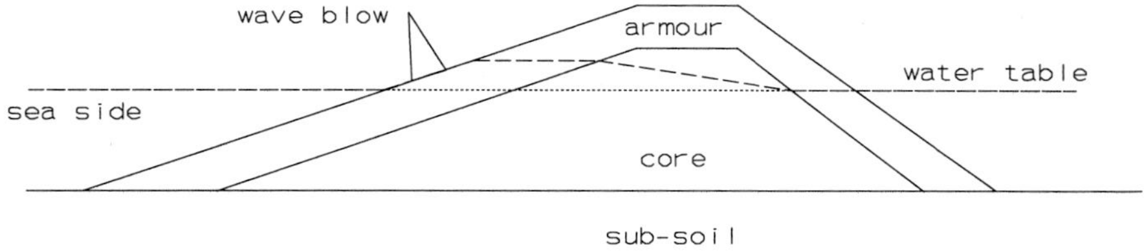

Figure 6.5 Cross-section of breakwater

6.3.3 Numerical model

Figure 6.6 shows the finite element mesh of the breakwater. Both the core and the armour are divided into two regions: a 'wet' region, under the water table, and a 'dry' region, above the water table. The model is plane strain.

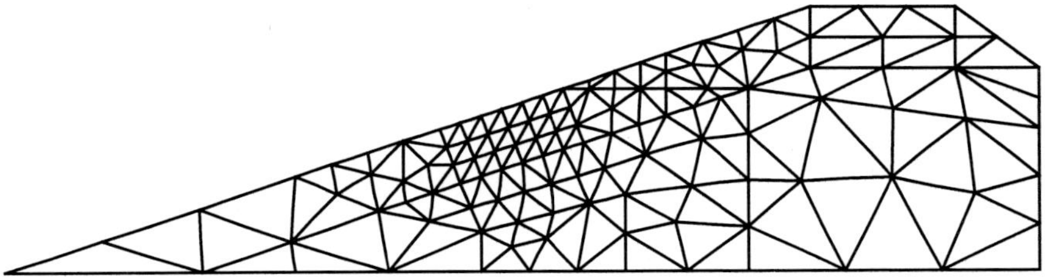

Figure 6.6 Division in elements of the breakwater

The material properties are shown in Table 6.3. The behaviour of the solid skeleton is assumed to be linear elastic. The bulk modulus of the water is chosen relatively low, since some influence of air bubbles might be expected.

The breaking wave loads the breakwater at the wet zone of the armour, just between the still water level and the internal water table. The wave is also shown in Figure 6.5. The width of the loaded area is 2 m, the distribution along the slope is triangular. The distribution in time of the load is also triangular: a increasing load during 20 ms, and a decreasing load during 60 ms. So, the duration of the loading is 80 ms. The maximum force is 150 kN. The loading is distributed over the solid and the fluid as the relative surface of the material (thus 40% on the fluid and 60% on the solid).

6.3.4 Breakwater on a stiff impermeable sub-soil

Now, the dynamical stresses in the solid and the fluid due to the breaking wave are calculated. The fluid stress, the first and second stress invariant of the solid stress, the displacement and the velocity of the mixture and the relative discharge are presented in Figure 6.7 (a to f) at several moments at 20 ms intervals: 20, 40, 60, 80, 100, 120, 140

Table 6.3 Properties of materials in the breakwater

variable	symbol	unit	core	armour	sand
volumetric mass fluid	ρ_f	kg/m^3	1000	1000	1000
volumetric mass solid	ρ_s	kg/m^3	2650	2650	2650
apparent mass	ρ_a	kg/m^3	200	175	-
porosity	n	-	0.35	0.40	0.35
bulk modulus fluid	K_f	MPa	200	200	200
bulk modulus solid	K_s	MPa	50	80	10
shear modulus solid	G_s	MPa	30	35	6
friction factor permeability	b k	kg/sm^3 m/s	12.0*10^3 0.10	3.1*10^3 0.51	-
equivalent celerity	c_{eq}	m/s	565	561	374

and 160 ms etc. In order to interpret the results, one must bear in mind that the mechanical sign definition for the stresses is used (compression is negative) and that only dynamical stresses due to the wave blow are shown.

From the contour lines of the deviator stress, it can be seen that a high deviator stress is generated in the middle under the load at the interface between core and armour. The value is about 25 kPa, at t = 40 ms. The influence of the layer separation on the solid stresses can be seen. This is in accordance with the discussion in chapter 3.

At this moment, the dynamic isotropic stress (mean stress) decreases rapidly with depth, whereas the deviator stress (shear stress) decreases very slowly. At 4 m under the slope, the vertical stress is about 30 kPa, so the stress ratio (deviator stress / isotropic stress) is 0.8, if the initial stress is isotropic.

At t = 80 ms a positive isotropic stress of 5 kPa is found, which means that tension occurs. However, the vertical stress due to the underwater weight of the solid is still 5 times bigger.

Comparison of the velocity and specific discharge plots, shows that the complete breakwater is loaded by the wave blow. However, the specific discharge has a non-zero value in the armour close to the location of the wave blow. Outside this region, the water and the skeleton move more or less together. The energy of the wave blow is thus dissipated by two mechanisms: firstly the water flow in the loaded area, and secondly the radiation of wave into the breakwater.

Another interesting phenomenon is occurrence of two waves in the breakwater. The compressional wave can be seen at the crest of the breakwater at 60 and 80 ms. The propagation velocity is approximately the equivalent value of 560 m/s (see Table 6.3). Afterwards, a strong downward velocity propagates along the crest. This is the Rayleigh wave. The theoretical Rayleigh wave velocity is approximately 140 m/s, or about 3 m per 20 ms. This value can be seen in the Figures.

At t = 160 ms, the large value of the horizontal velocity in the toe of the breakwater is remarkable. In reality this velocity might be smaller, since some energy will be dissipated in the water above the wet part of the slope. However, the radiation into the subsoil will lead to shear stress in the interface between the breakwater and the subsoil. This can be seen in the Figure for the deviator stress at t = 180 ms.

6.3.5 Breakwater on a sandy subsoil

In order to get some insight in the influence of a weaker subsoil, the boundary condition at the bottom has been changed. The sandy subsoil is modelled by a transmitting boundary. The stiffness of the sandy subsoil is one-fifth if the stiffness of the core.
The results are presented in the Figure 6.8 (a to f). These Figures are similar figures as before, but only at t = 20, 40 60 and 80 ms. The figures for the velocity and the figures for the specific discharge are continued until 200 ms.

From the figures for the stresses, the differences are seen at the bottom of the breakwater, as could be expected.

Comparison of the velocity-vectors shows that considerable velocities occur at the bottom of the breakwater. These velocities at the bottom are quite local. At the beginning it is due to the compressional wave under the loaded area, later it is due to the propagation of the Rayleigh wave.

The horizontal velocity at the toe of the breakwater occurs again, but is smaller than the case of a infinitely stiff rock at the bottom.

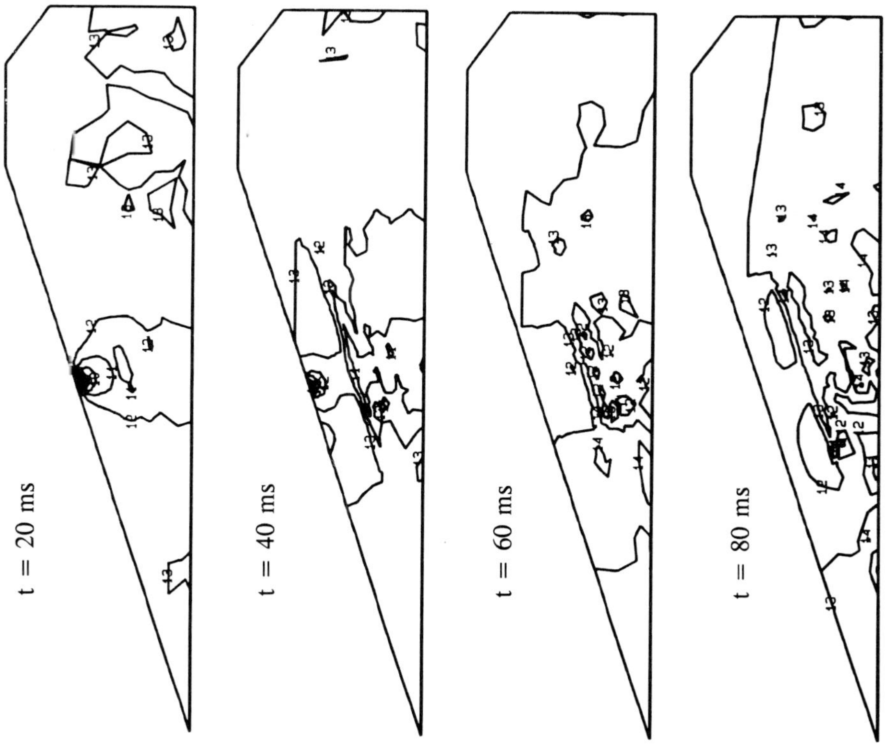

Figure 6.7.a Breakwater on stiff impermeable subsoil, fluid stress

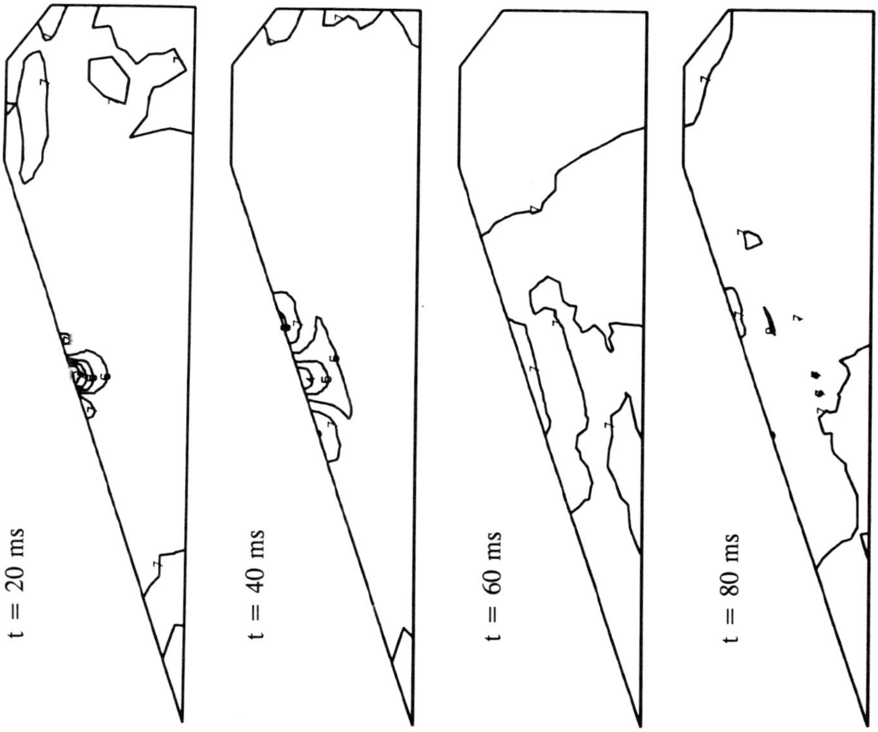

Figure 6.7.b Breakwater on stiff impermeable subsoil, mean solid stress

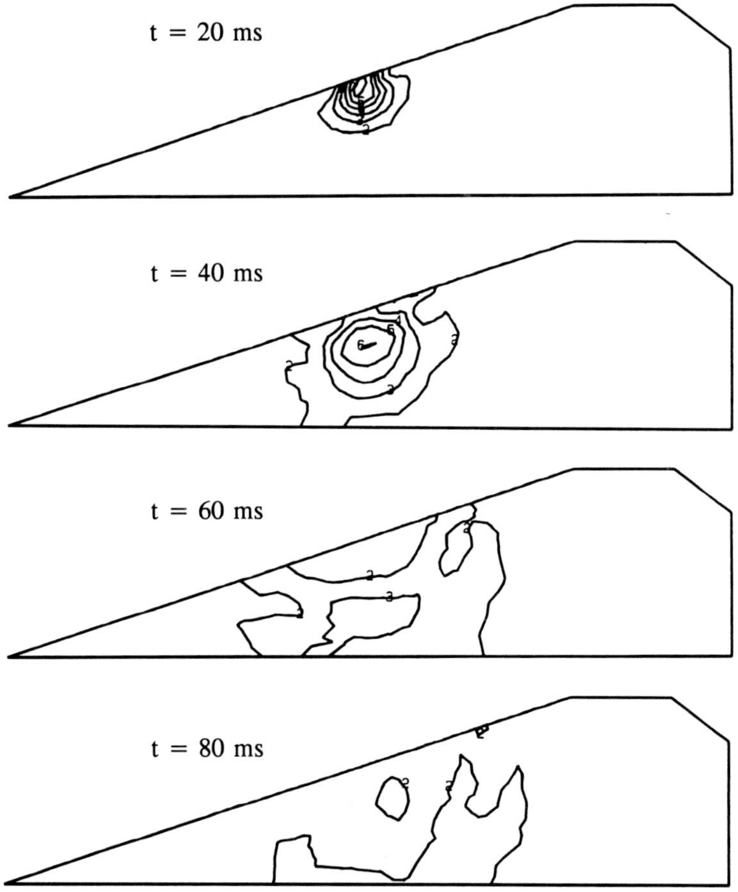

Figure 6.7.c *Breakwater on stiff impermeable subsoil, deviator stress*

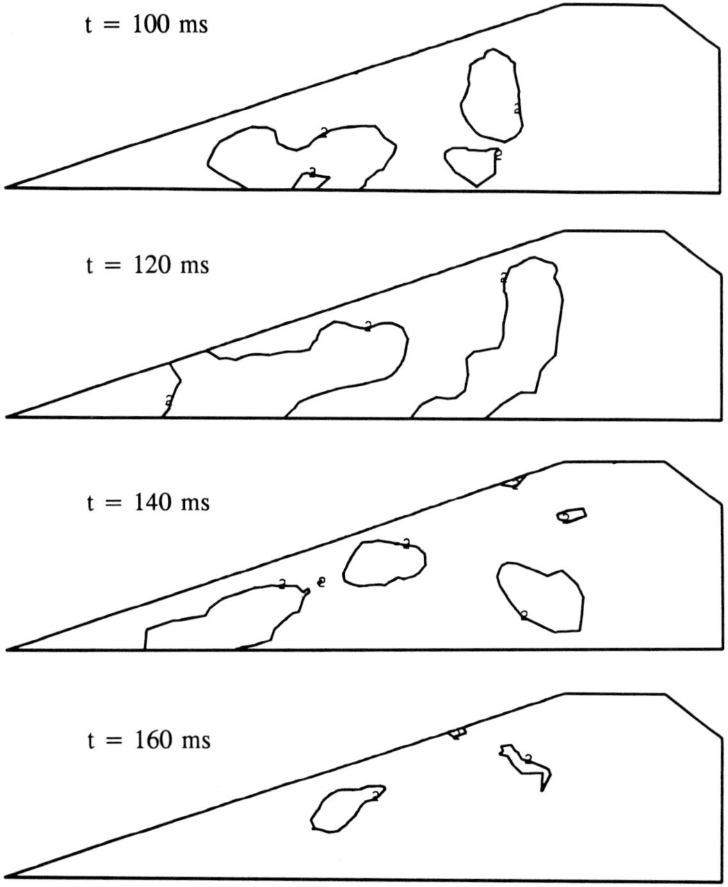

Figure 6.7.c *Breakwater on stiff impermeable subsoil, deviator stress (cont.)*

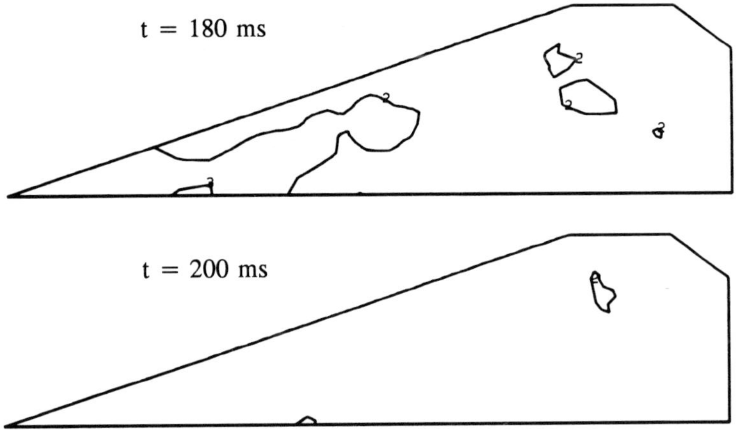

Figure 6.7.c *Breakwater on stiff impermeable subsoil, deviator stress (cont.)*

Legend for Figure 6.7 a, b and c values in kPa

line	fluid	line	solid	deviator
1	-60	1	-30	0
3	-50	2	-25	5
5	-40	3	-20	10
7	-30	4	-15	15
9	-20	5	-10	20
11	-10	6	-5	25
13	0	7	0	30
15	10	8	5	
		9	10	

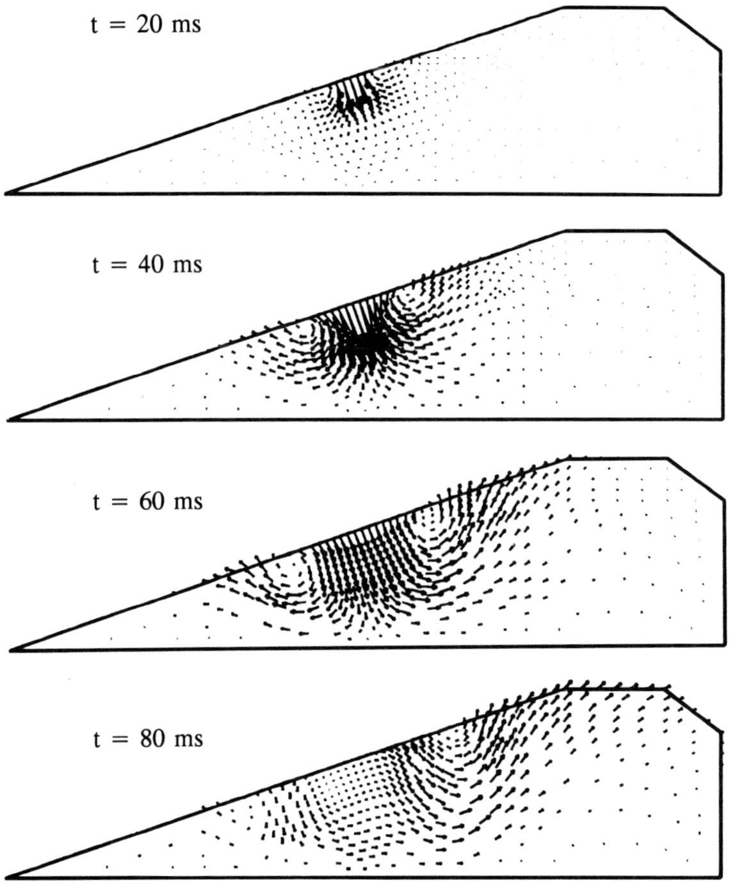

Figure 6.7.d *Breakwater on stiff impermeable subsoil, displacements*

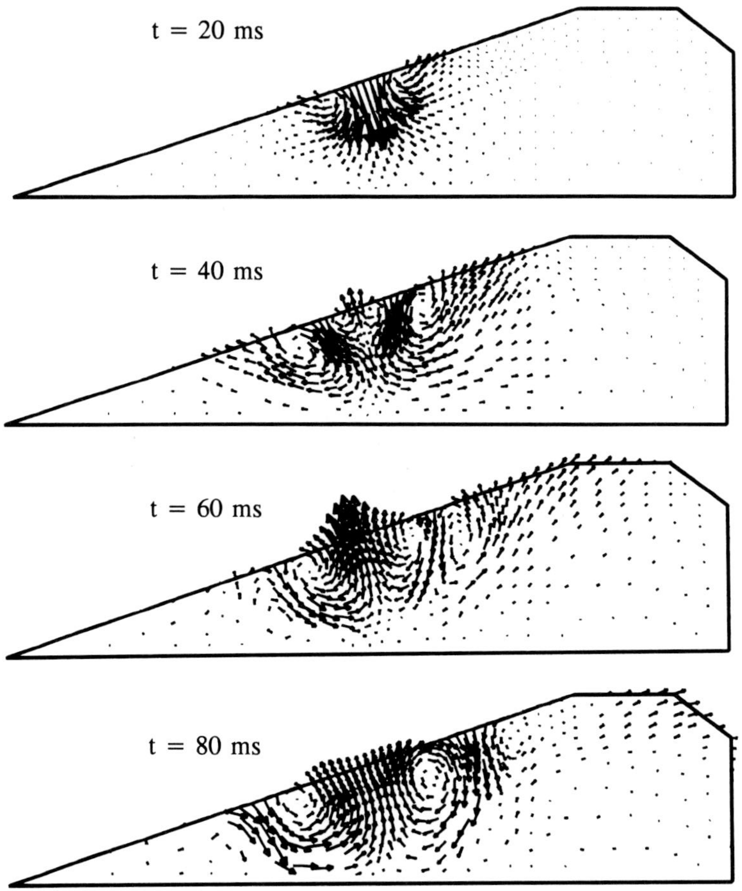

Figure 6.7.e Breakwater on stiff impermeable subsoil, velocity

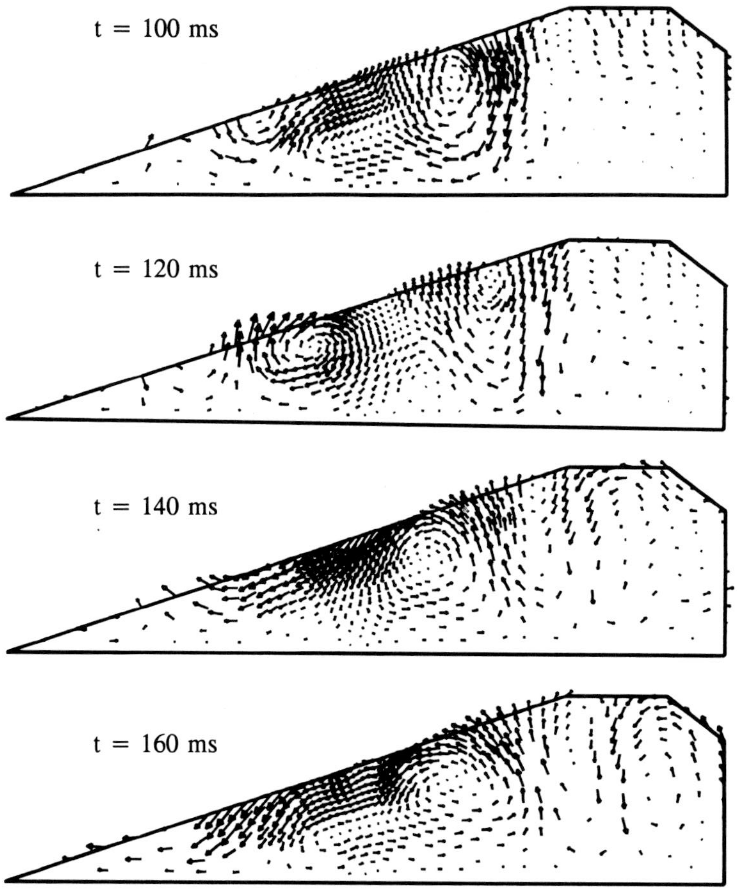

Figure 6.7.e Breakwater on stiff impermeable subsoil, velocity (cont.)

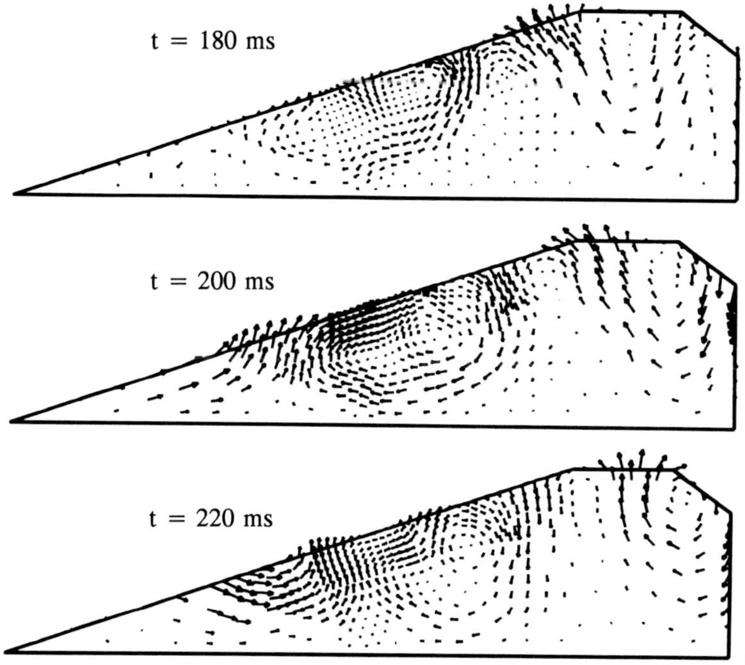

Figure 6.7.e *Breakwater on stiff impermeable subsoil, velocity (cont.)*

Legend for Figure 6.7 d, e and f

Exaggeration factors:

Solid displacement	1000
Solid velocity	50
Specific discharge	20

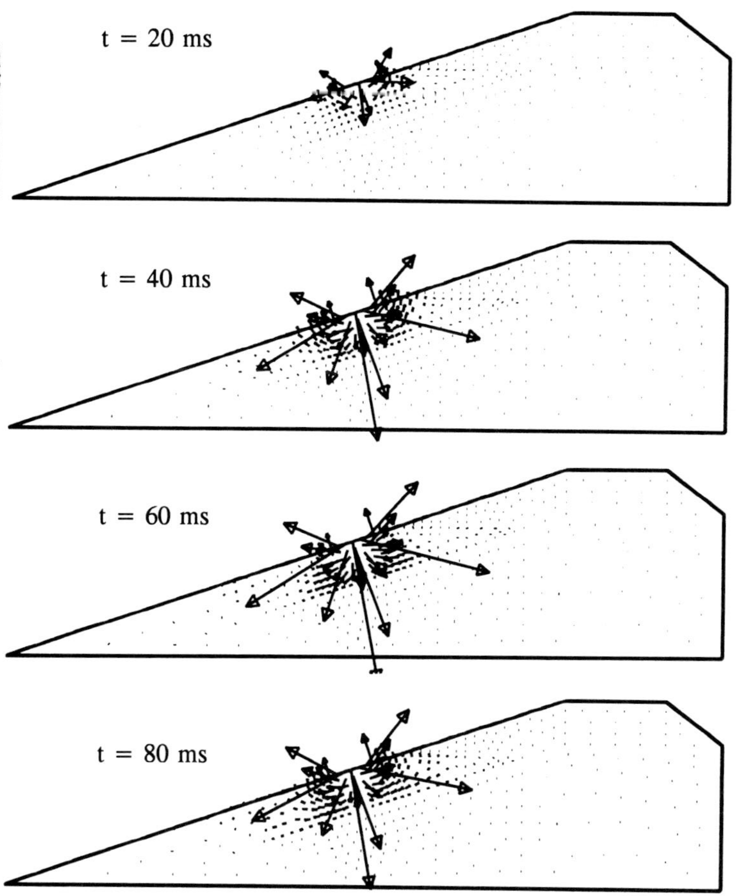

Figure 6.7.f *Breakwater on stiff impermeable subsoil, specific discharge*

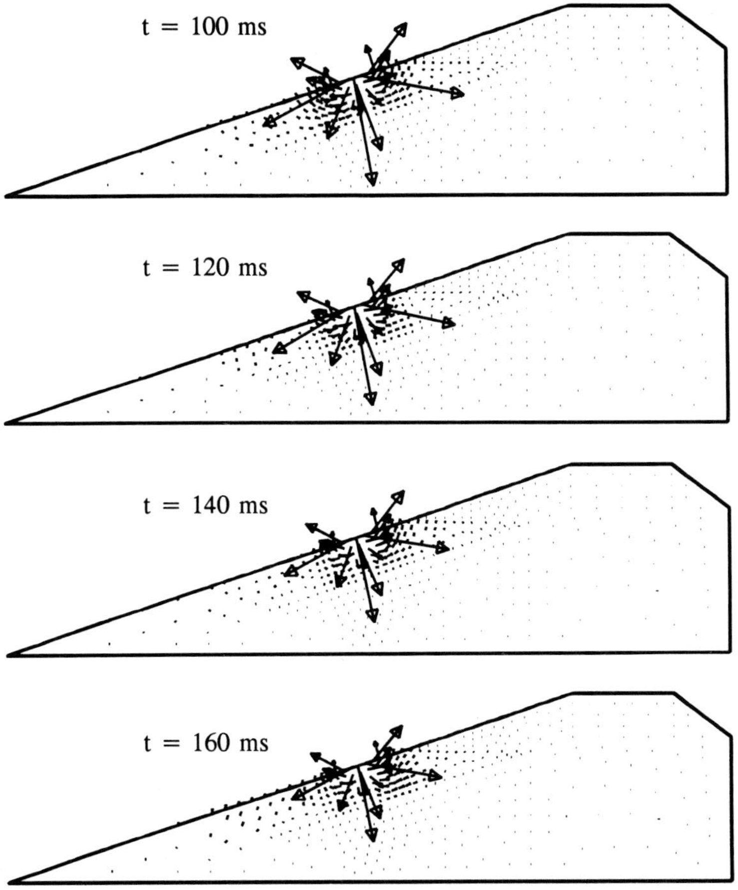

Figure 6.7.f *Breakwater on stiff impermeable subsoil, specific discharge (cont.)*

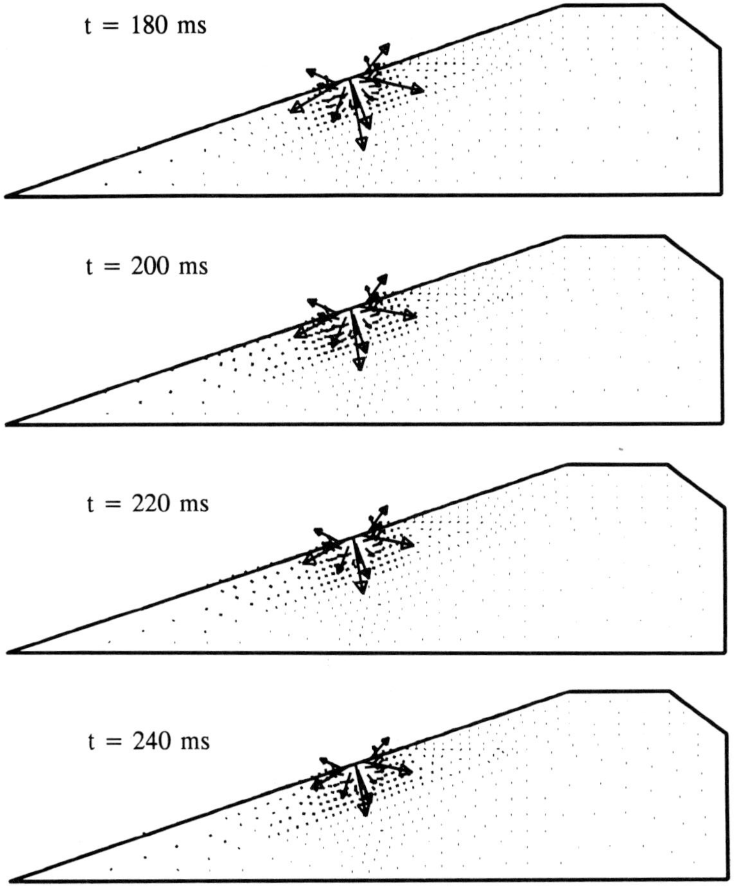

Figure 6.7.f *Breakwater on stiff impermeable subsoil, specific discharge (cont.)*

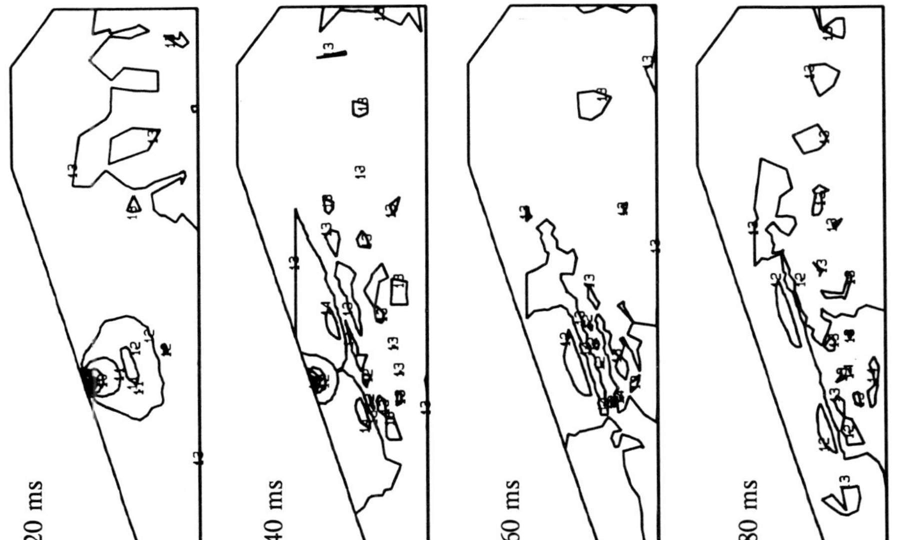

Figure 6.8.a Breakwater on sandy subsoil, fluid stress

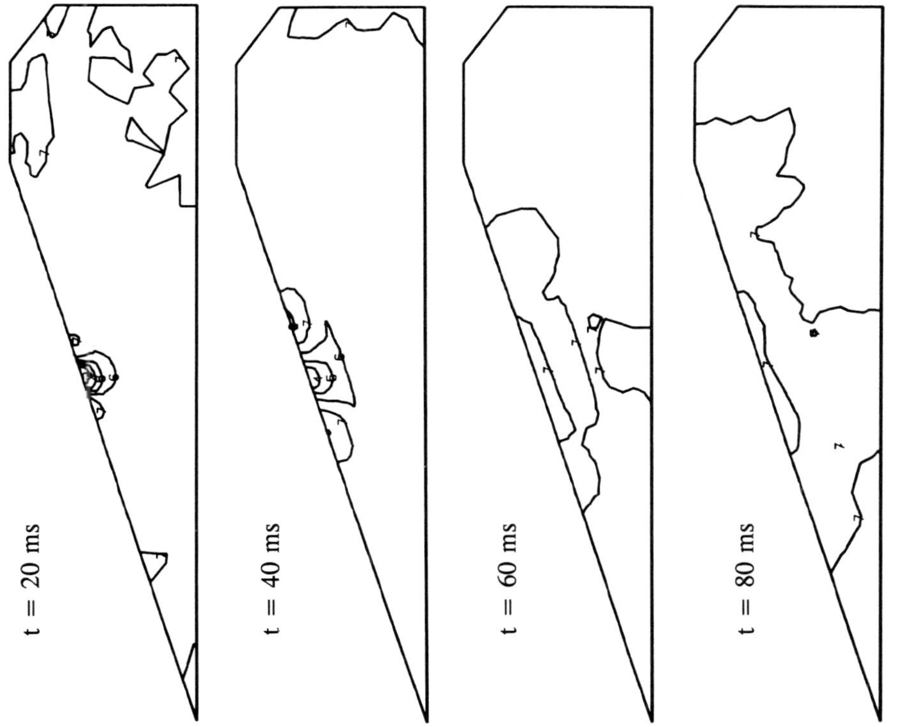

Figure 6.8.b Breakwater on sandy subsoil, mean solid stress

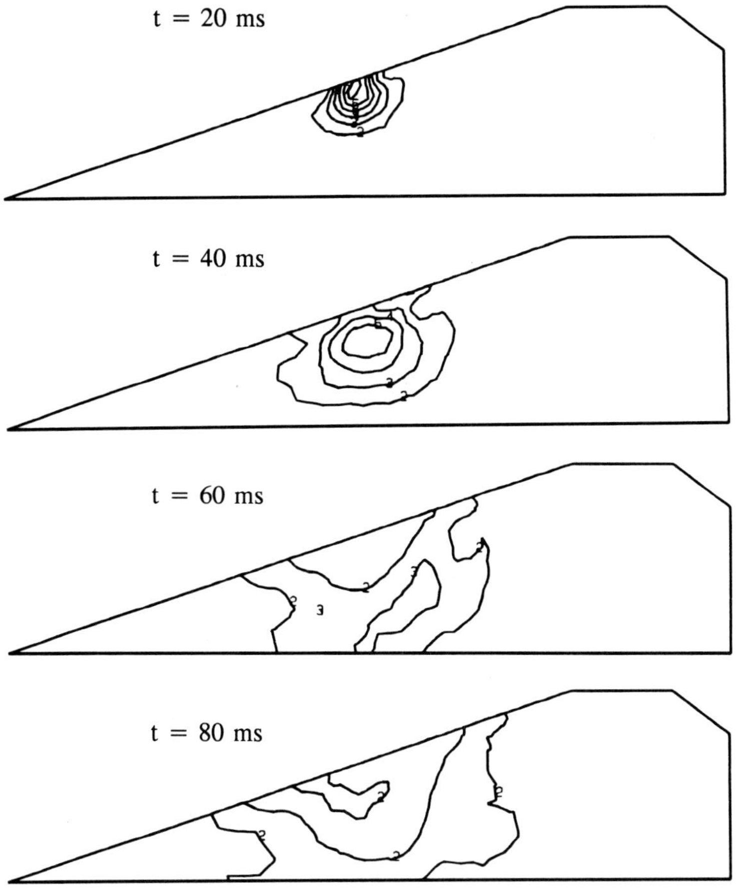

Figure 6.8.c   Breakwater on sandy subsoil, deviator stress

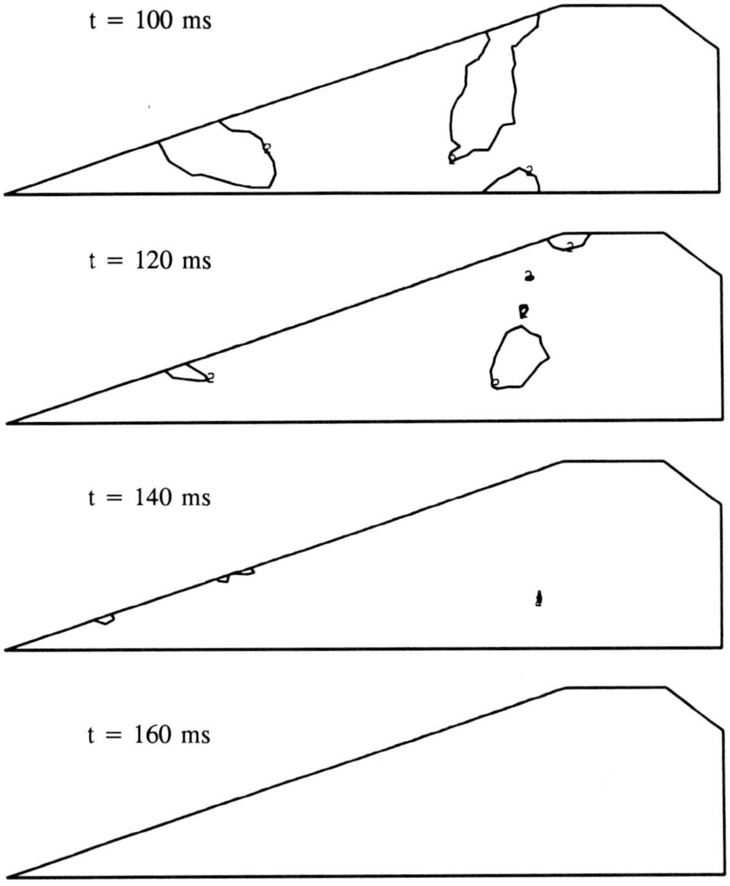

Figure 6.8.c   Breakwater on sandy subsoil, deviator stress (cont.)

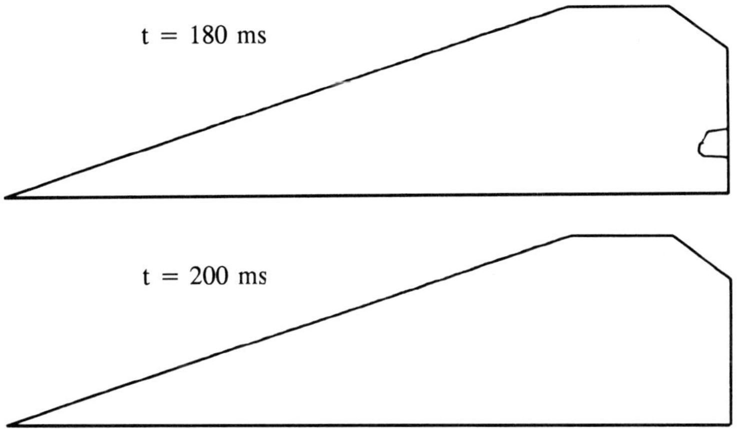

Figure 6.8.c *Breakwater on sandy subsoil, deviator stress (cont.)*

Legend for Figure 6.8 a, b and c values in kPa

line	fluid	line	solid	deviator
1	-60	1	-30	0
3	-50	2	-25	5
5	-40	3	-20	10
7	-30	4	-15	15
9	-20	5	-10	20
11	-10	6	-5	25
13	0	7	0	30
15	10	8	5	
		9	10	

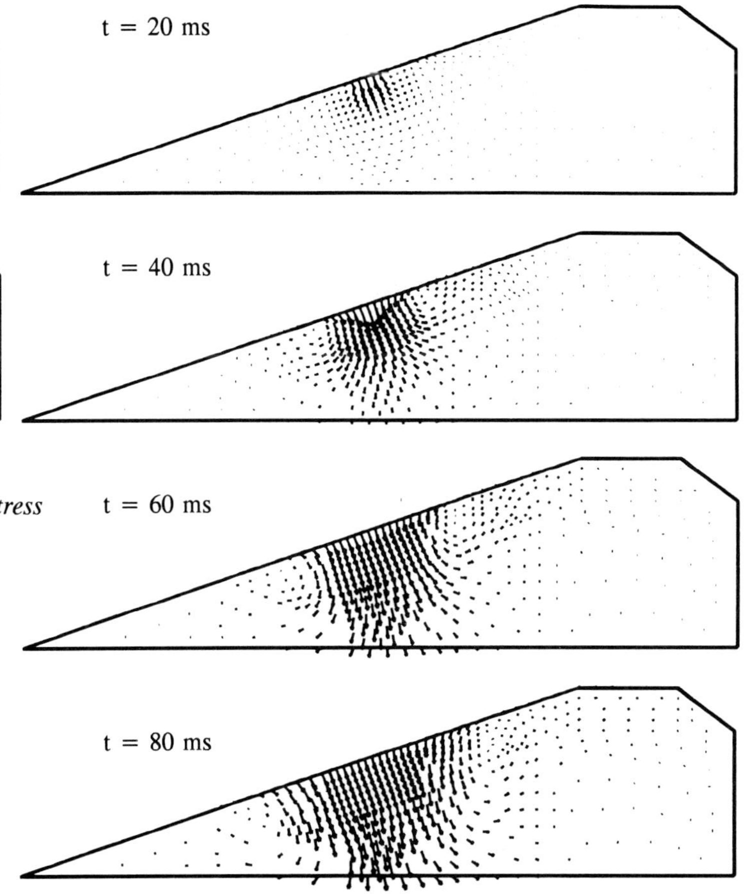

Figure 6.8.d *Breakwater on sandy subsoil, displacements*

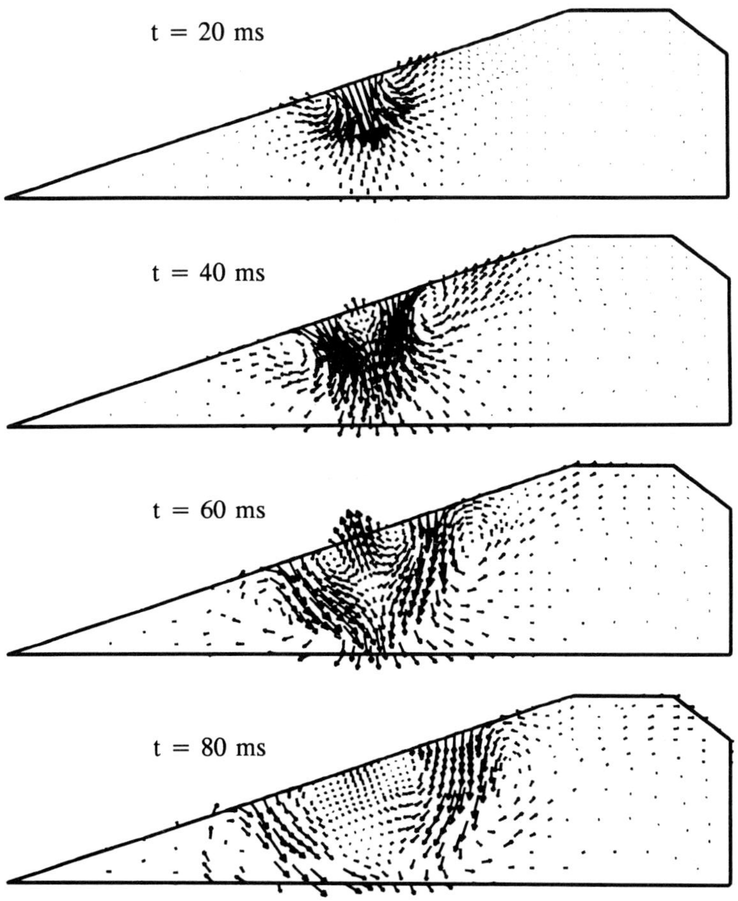

Figure 6.8.e Breakwater on sandy subsoil, velocity

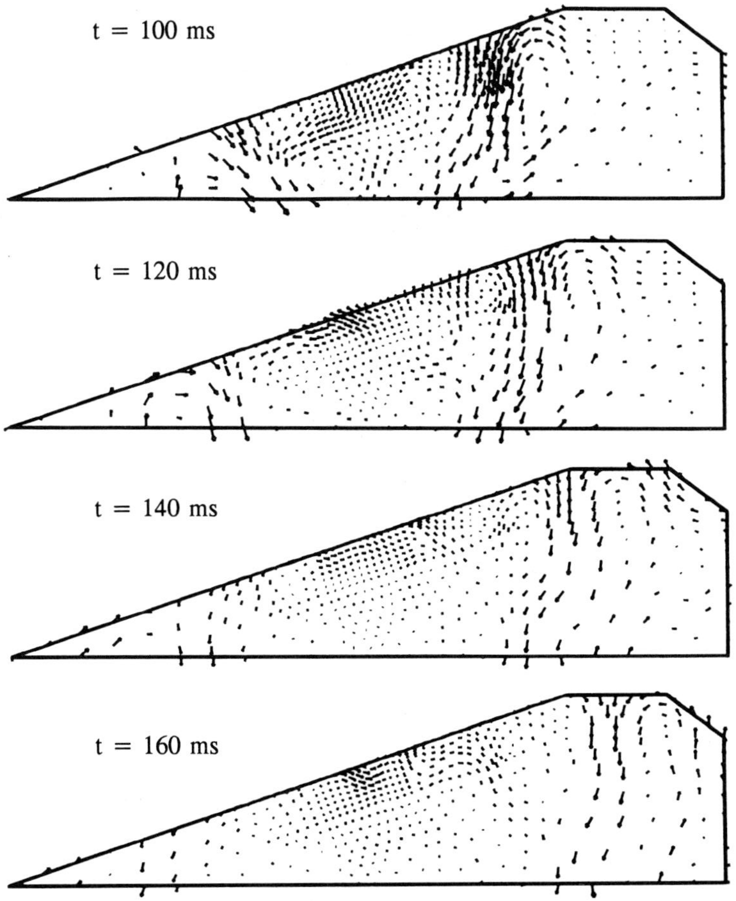

Figure 6.8.e Breakwater on sandy subsoil, velocity (cont.)

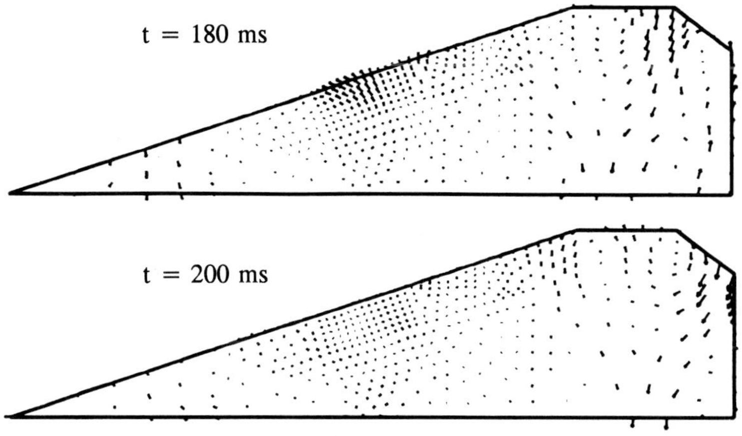

Figure 6.8.e Breakwater on sandy subsoil, velocity (cont.)

Legend for Figure 6.8 d, e and f

Exaggeration factors:

Solid displacement	2000
Solid velocity	50
Specific discharge	20

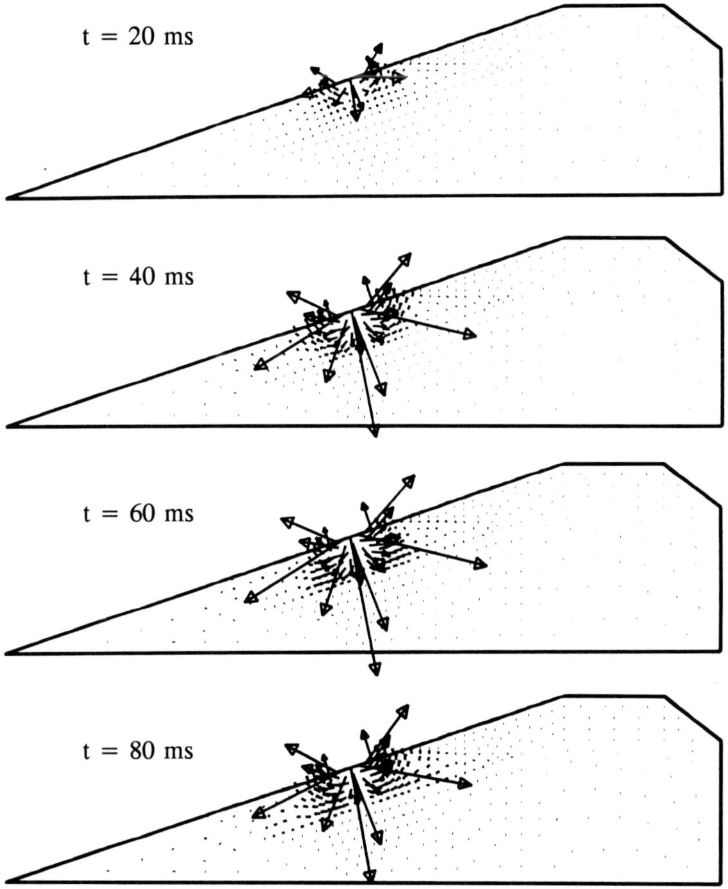

Figure 6.8.f Breakwater on sandy subsoil, specific discharge

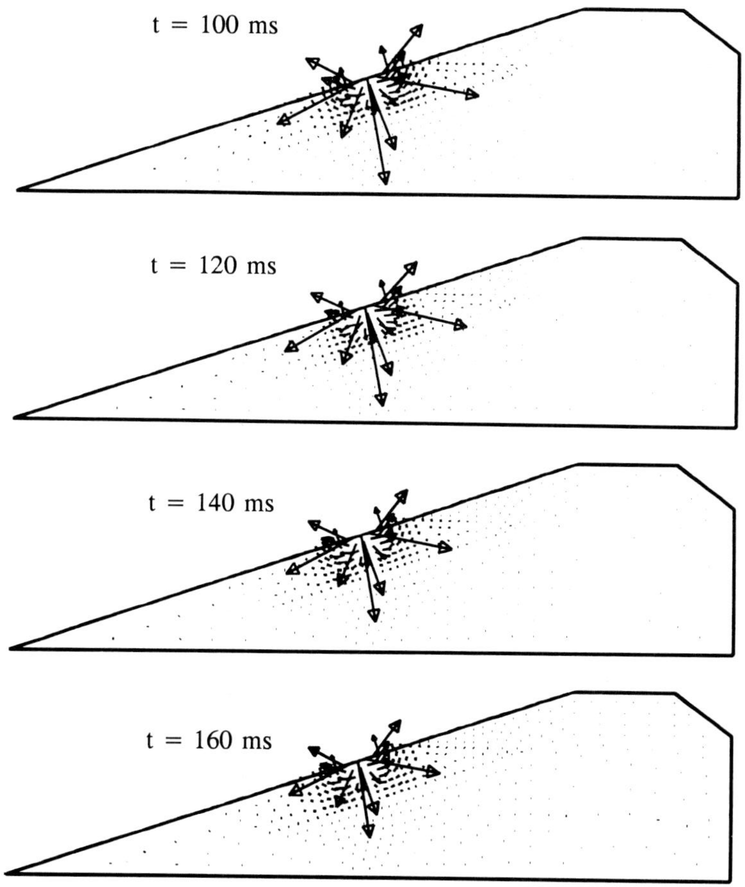

Figure 6.8.f *Breakwater on sandy subsoil, specific discharge (cont.)*

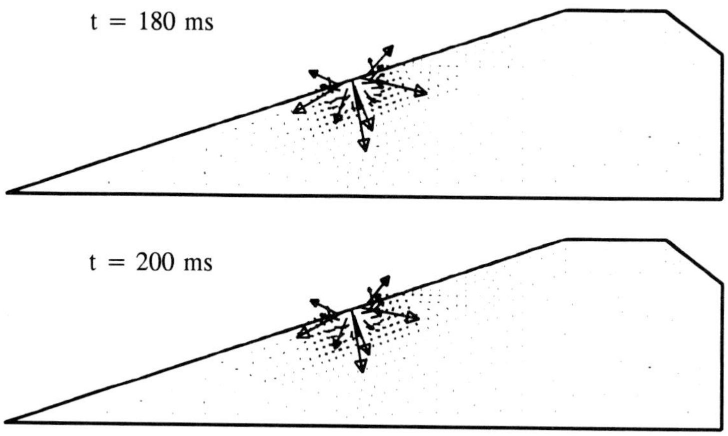

Figure 6.8.f *Breakwater on sandy subsoil, specific discharge (cont.)*

7 CONCLUSIONS

Close to an interface, the solid stress caused by a compressional wave in a saturated porous material is influenced by the wave of the second kind. For the practical values of the permeability and the stiffness of the solid, this may lead to a decrease in the solid stress of up to 70%. This cannot be described by a single phase approach.

When the behaviour of the shear wave in a saturated medium is compared to the behaviour of the shear wave in the equivalent single phase material, only minor differences are to be expected.

The existence of the second compressional wave does not lead to two Rayleigh waves at the surface of a saturated porous half-space. The two-phase behaviour leads to a frequency-dependent damping of the Rayleigh wave.

The field equations which describe the behaviour of a saturated porous medium can be rewritten to a description in which the solid displacement, fluid stress and specific discharge appear as variables. After discretisation according to the standard finite element method, the specific discharge is eliminated by using dynamic condensation, while time integration is effected by Heun's method. The resulting calculation scheme is explicit in time. This scheme is suitable for the simulation of both fast and slow phenomena. Allowance can be made for the apparent mass of the saturated porous medium.

Using this method of discretisation as a basis, the finite element program Titan is developed and tested. The non-linear behaviour of the solid is described using an elastoplastic non-linear hardening model for the case of low mean stress, and a bilinear elastic model for the case of high mean stress.

Three dynamical experiments are carried out:

- pile-driving simulator
- full-scale driving test
- photoelastic test

The pile-driving simulator offers an insight into the wave radiation under a pile toe in a small sample. In general, the energy radiates into the soil under the pile toe in a fairly narrow cone. The area of radiation in saturated sand is smaller than in dry sand.

The limited area of radiation is confirmed in the full-scale field test. The material velocity of the soil exhibits similar behaviour during driving and dynamic testing, but the maximum value and damping are different in each case. These measurements are processed to obtain a complete description of the velocity field around the pile toe during driving. During the quasi-static loading, dynamical effects and considerable fluid stresses are measured. This shows that the soil also behaves dynamically under quasi-static loading of the pile.

The dynamical solid (deviator) stresses in a saturated porous material can be measured in an oil-glass mixture. Accurate measurements can be obtained by using laser-light and a photo-diode, while an overall insight into the wave propagation can be acquired by the use of a high-speed camera. The area with high dynamical stresses is limited, as in a static test.

Application of the finite element program Titan to a breakwater under wave attack shows considerable extension of the stress waves due to the wave impact. The area in which fluid flows is rather limited. At interfaces between several layers, high deviator stresses occur. The energy dissipation caused by the fluid flow close to the impact cannot be described by a single phase approach.

The tests involving the pile-driving simulator are calculated with Titan. Similar differences to those found in the experiments for dense sand are also found when the linear model is used. This shows that, for loose sand, the non-linearity cannot be ignored.

REFERENCES

Allersma, H.G.B., 1987
Optical analysis of stress and strain in photo-elastic particle assemblies
Thesis, Delft University of Technology, Delft, the Netherlands
Allersma, H.G.B., 1991
Using image processing in centrifuge research
Proceedings International Conference on Centrifuges, ed. Ko, H.Y., McLean, F.G., Balkema, Rotterdam, the Netherlands, pp. 551-558
Alonso, E.E., Gens, A., Josa, A., 1990
A constitutive model for partially saturated soils
Geotechnique, Vol. 40, No. 3, September 1990, pp. 405-430
Biot, M.M., 1956
Theory of propagation of elastic waves in fluid saturated porous solid, I. Low frequency range
The Journal of the Acoustical Society of America, Vol. 28, No. 2, March 1956, pp. 168-178
Biot, M.M., 1962
Mechanics of deformations and acoustic propagation in porous media
Journal of applied physics, Vol. 33, No. 4, April 1962, pp. 1482-1498
Bourbie, T., Coussy, O., Zinszner, B., 1987
Acoustics of porous media
Edition Technip, Paris, France
Brinkgreve, R.B.J., 1994
Geomaterial models and numerical analysis of softening
Thesis, Delft University of Technology, Delft, the Netherlands
Chavent, G., Roberts, J.E., 1991
A unified physical presentation of mixed, mixed-hybrid finite elements and standard finite difference approximations for the determination of velocities in waterflow problems
Advances in Water Resources, Vol. 14, No. 6, pp. 329-348
Degrande, G, 1992
A spectral and finite element method for wave propagation in dry and saturated poroelastic media
Thesis, Catholic University Leuven, Leuven, Belgium
Deresiewicz, H., 1960
The effect of boundaries on wave propagation in a liquid-filled porous solid: I. Reflection of plane waves at a free plane boundary (non-dissipative case)
Bulletin of the Seismological Society of America, Vol. 50, No. 4, October 1960, pp. 783-788
Deresiewicz, H., 1960
The effect of boundaries on wave propagation in a liquid-filled porous solid: IV. Surface waves in a halfspace
Bulletin of the Seismological Society of America, Vol. 50, No. 3, July 1960, pp. 627-638

Deresiewicz, H. & Skalak, R., 1963
On uniqueness in dynamic poroelasticity
Bulletin of the Seismological Society of America, Vol. 53, No. 4, July 1963, pp. 783-788

Geerling, J., Smits, M.Th.J,H., 1995
Testbook of the Measurements and Contest at the fourth International Conference on the Application of Stress Wave Theory on Piles, The Hague, 1992, to be published by Balkema, Rotterdam.

Geertsma, J. & Smit, D.C., 1961
Some aspects of elastic wave propagation in fluid-saturated porous solids
Geophysics, Vol. 26, No. 2, April 1961, pp. 169-181

Ghaboussi, J., Wilson, E.L., 1972
Variational formulation of dynamics of fluid-saturated porous elastic solids
Proceedings of the American Society of Civil Engineers, Journal of the Engineering Mechanics Division, Vol. 98, No. EM4, August, 1972, pp. 947-963

Hilber, H.M., Hughes, T.J.R., Taylor, R.L.
Improved numerical dissipation for time integration algorithms in structural dynamics
Earthquake Engineering and Structural Dynamics, Vol. 5, 1977, pp. 283-292

Hölscher, P., 1992
Numerical simulation of a fluid-saturated soil under a dynamically loaded pile
Transport in Porous media, ed. Bear., J., Vol. 9, No. 1&2, October 1992, pp. 73-84

Hölscher, P., Barends, F.B.J., 1993
Mixed finite element method for dynamic analysis of fluid-saturated porous media
Proceedings of the Second European Conference on Structural Dynamics: Eurodyn'93, Trondheim, Norway, 21-23 june, 1993, pp. 381-386
A.A. Balkema, Rotterdam , the Netherlands, 1993

Hughes, T.J.R., 1987
The Finite Element Method
Prentice-Hall Inc., Englewood Cliffs, New Jersey, USA

Ishihara, K., 1970
Approximate forms of wave equations for water-saturated porous materials and related dynamic modulus
Soils and foundations, Vol. 10, no. 4, 1970, pp. 10-38

Ishihara, K. & Okada, S., 1978
Yielding of overconsolidated sand and liquefaction model under cyclic stress
Soils and Foundations (Japanese Society of Soil Mechanics and Foundation Engineering), Vol. 18, No. 1, March 1978, pp. 57-72

Janes, M.C., Horvath, R.C., Bermingham, P.D., 1991
An Innovative Dynamic Test Method for Piles
Proceedings Second International Conference on Recent Advances in Geotechnical Earthquake Engineering and Soil Dynamics, March, 1991, St.Louis, MO., University of Rolla, MO., Prakash, S. (ed), Vol. I, pp. 253-261, 1991

Kaasschieter, E.F., Huijben, A.J.M., 1992
Mixed-hybrid finite elements and streamline computations for the potential flow problem
Numerical Methods for Partial Differential Equations, Vol. 8, pp. 221-266

Oka, F., 1992
A cyclic elasto-viscoplastic constitutive model for clay based on the non-linear hardening rule
Proceedings of the 4th International Symposium on Numerical Models in Geomechanics NUMOG IV, eds. Pande, G.N. & Pietruszczak, S., Swansea, 24-27 August 1992, Balkema, Rotterdam, the Netherlands, pp. 105-114

Oka, F., Yashima, A., Shibata, T., Kato, M., Uzuoka, R., 1994
FEM-FDM coupled liquefaction analysis of an porous soil using an elasto-plastic model, to be published

Park, K.C., Felippa, C.A., 1983
Partitioned Analysis of Coupled Systems
Chapter 3 in 'Computational Methods in Transient Analysis', ed. Belytschko, T., Hughes, T.J.R., Elsevier Science Publishers B.V., the Netherlands, pp. 157-219

Park, K.C., 1984
Stabilization of partitioned solution procedure for pore fluid-soil interaction analysis
International Journal for Numerical Methods in Engineering, Vol. 19, pp. 1669-1673

Plona, T.J., 1980
Observation of a second bulk compressional wave in a porous medium at ultrasonic frequencies
Applied Physical Letters, 1980, Vol. 36, no. 4, pp. 259-261

Polytechnisch zakboekje PBNA, 1980
38th print, Arnhem, the Netherlands, 1980

Prevost, J.H., 1985
Wave propagation in fluid-saturated porous media: An efficient finite element procedure
Soil Dynamics and Earthquake Engineering, 1985, Vol. 4, No. 4, pp. 183-202

Raviart, P.A., Thomas, J.M., 1977
A mixed finite element method for 2nd-order elliptic problems
in Mathematical aspects of finite element methods, Lecture notes in mathematics, Vol. 606, Galligini, I., Magenes, E. eds., Springer, Berlin, 1977, pp. 292-315

Rossmanith, H.P., Shukla, A., 1982
Photoelastic investigation of dynamic load transfer in granular media
Acta Mechanica, 1982, No. 42, pp. 211-225

Sandhu, R.S., Shaw, H.L., Hong, S.J., 1990
A three-field finite element procedure for analysis of elastic wave propagation through fluid-saturated soils
Soil Dynamics and Earthquake Engineering, 1990, Vol. 9, No. 2, pp. 58-65

Shell, undated
Shell smeermiddelen voor industrie en scheepvaart
Shell Nederland Verkoopmaatschappij B.V., Rotterdam, the Netherlands

Shukla, A., Damania, C., 1987
Experimental investigation of wave velocity and dynamic contact stresses in an assembly of disks
Experimental Mechanics, September 1987, pp. 268-281

Shukla, A., Zhu, C.Y., Sadd, M., 1988
Angular dependence of dynamic load transfer due to explosive loading in granular aggregate chains
Journal of Strain Analysis, Vol. 23, No. 3, 1988, pp. 121-127
Shukla, A.,1991
Dynamic photoelastic studies of wave propagation in granular media
Optics and Lasers in Engineering, 1991, Vol. 14, pp. 165-184
Simon, B.R., Wu, J.S.S., Zienkiewicz, O.C., 1986
Evaluation of higher order, mixed and Hermitean finite element procedures for dynamic analysis of saturated porous media using one-dimensional models
International Journal for Numerical and Analytical methods in Geomechanics, 1986, Vol. 10, pp. 483-499
Simon, B.R., Wu, J.S.S., Zienkiewicz, O.C., Paul, D.K., 1986
Evaluation of u-w and u-π finite element methods for the dynamic analysis of saturated porous media using one-dimensional models
International Journal for Numerical and Analytical methods in Geomechanics, 1986, Vol. 10, pp. 461-482
Simon, B.R., Zienkiewicz, O.C., Paul, D.K., 1984
An analytical solution for the transient response of saturated porous elastic solids
International Journal for Numerical and Analytical Methods in Geomechanics, 1984, Vol. 8, pp. 381-398
Smeulders, D.M.J., 1992a
On wave propagation in saturated and partially saturated porous media
Thesis, Eindhoven University of Technology, Eindhoven, the Netherlands
Smeulders, D.M.J., 1992b
private communications
Stuit, H., 1992
private communications
Sweet, J.S., Barends, F.B.J., Van Loon Engels, C., Van der Kogel, H., 1980
A method for dynamical soil-structure interaction problems
Proceedings of the International Conference on Numerical Methods for Non-linear Problems, eds. Taylor, C., Hinton, E., Owen, D.R.J., Swansea, 1980, pp. 197-207
Tajuddin, M., 1984
Rayleigh waves in a poroelastic halfspace
Journal of the Acoustic Society of America, Vol. 75, No. 3, March, 1984, pp. 682-684
Trense, R.W., 1970
Schokgolven in granulaire substanties
De Ingenieur, Vol. 82, No. 45, November 1970, pp. 137-146
Van der Graaf, L, 1964
Foto-elastische onderzoek naar het gedrag van granulair materiaal dat in diverse graden van verzadiging verkeert
Verslag no. 103 B, Advies bureau der genie, the Netherlands, 1964
Van der Grinten, J.G.M., Van Dongen, M.E.H., Van der Kogel, H., 1985
A shock-tube technique for studying pore-pressure propagation in a dry and saturated porous medium
Journal of Applied Physics, Vol. 58, No. 8, october, 1985, pp. 2937-2942

Van der Grinten, J.G.M., Van Dongen, M.E.H., Van der Kogel, H., 1987
Strain and pore pressure propagation in a water-saturated porous medium
Journal of Applied Physics, Vol. 62, No. 12, december, 1987, pp. 4682-4687

Van der Grinten, J.G.M., 1987
An experimental study of Shock-induced wave propagation in dry, water-saturated, and partially saturated porous media
Thesis, Eindhoven University of Technology, Eindhoven, the Netherlands

Van der Kogel, H., 1977
Wave propagation in saturated porous media
Thesis, California Institute of Technology, Pasadena, California, USA

Van Kan, J.J.L.M., 1992
private communications

Verruijt, A., 1982
Some remarks on the principle of effective stress
In: Vermeer, P.A., Luger, H.J. eds., Proc. of the symp. on deformation and failure of granular materials, Delft, August 31-September 3, 1982, pp. 167-170
A.A. Balkema, Rotterdam, the Netherlands, 1982

Zienkiewicz, O.C., Bettes, P., 1982
Soils and other saturated media under transient dynamic conditions: General formulation and the validity of various assumptions
in: Pande, G.N., Zienkiewicz, O.C. eds., Soil Mechanics - Transient and cyclic loads, John Wiley and Sons, 1982, pp. 1-16

Zienkiewicz, O.C., Paul, D.K., Chan, A.H.C., 1988
Unconditionally stable staggered solution procedure for soil-pore fluid interaction problems
International Journal for Numerical Methods in Engineering, 1988, Vol. 26, pp. 1039-1055

Zienkiewicz, O.C., Taylor, R.L., 1989
The Finite Element Method, Vol I
McGraw-Hill Book Company ltd., London, UK

LIST OF SYMBOLS

In general all symbols are explained in detail in the text.

Latin

A	*	Multiplier
	-	Material constant non-linear model
	m	Amplitude
b	$kg/(sm^3)$	friction coefficient ($b=n^2\rho_f g/k$) unit also $N/m^3/(m/s)$
B	-	B-matrix
	*	Multiplier
	-	Material constant non-linear model
C	*	Multiplier
	*	Two subscripts: damping matrix (solid part) or mass matrix (fluid part)
c	m/s	Celerity of wave (wave velocity)
D	N/m^2	Stiffness matrix or tensor
e	-	Void ratio
f	*	One subscript: right hand side vector
	-	(Yield) surface non-linear model
	Hz	Frequency
	*	General function
g	m/s^2	Acceleration due to gravity (9.81 m/s^2)
	*	General function
	-	Plastic potential
G	Pa	Shear modulus
k	m/s	Hydraulic permeability
K	m^2	No subscript: intrinsic permeability
	Pa	One subscript: bulk modulus
	*	Two subscripts: stiffness matrix
K_c	Pa	Constrained modulus of skeleton
L	m	Length
m	kg/m^3	One or two subscripts: mass factors
M	*	Two subscripts: mass matrix
	-	Material constant non-linear model
	N/m^2	Stiffness fluid with compressibility of grains
n	-	Porosity of the solid
	-	Time step in time discretisation
N	-	Shape function
q	m/s	Specific discharge
Q	*	Two subscripts: inverse matrix for discharge system
R	N/m^3	Volumetric friction of fluid on solid
s	N/m^2	Two subscripts: deviator stress
S	*	Two subscripts: stiffness matrix
t	s	Time
u	m	Displacement
v	m/s	Particle velocity
V	m^3	Volume
w	m	Specific displacement of fluid ($w_j = n(u_j^f - u_j^s)$)
x	m	No or one subscript: space coordinate
	-	Two subscripts: hardening tensor
Z	Ns/m	Mechanical impedance

Greek

α	-	Integration constant Hilber-Hughes-Taylor method
	-	Influence of stiffness grains
β	-	Integration constant Hilber-Hughes-Taylor method
γ	-	Integration constant Hilber-Hughes-Taylor method
	rad	Shear angle
δ_{ij}	-	Kronecker delta (=1 if i=j; =0 if i≠j; tensor rank two)
Δ#	-	Increment or change in variable
Δ	s	Numerical time step
ε	-	Strain (tensor)
η	-	Stress ratio (tensor)
κ	m^{-1}	Wave number
	-	Swelling index
κ_s	$m^2/s/kg$	Permeability of the interface
λ	N/m^2	First Lamé constant
	-	Compression index
μ	N/m^2	Second Lamé constant
ν	-	Poison ratio
π	Pa	Fluid stress (minus pore water pressure, scalar)
ρ	kg/m^3	Volumetric mass
σ	Pa	Stress (tensor)
τ	Pa	Shear stress
	s	Duration time
φ	-	Scalar potential
ψ	-	Vector potential
ω	rad/s	Angular frequency

superscripts and subscripts
(if not explained above)

1	Solid
	First space coordinate
2	Fluid
	Second space coordinate
a	Apparent (volumetric mass only)
	Material number
b	Material number
c	Critical ((angular) frequency)
	Constrained (solid modulus)
d	Dry
e	Elastic
eq	Equivalent value (cf. Sec. 2.2.4)
f	Fluid
g	Grain
i	Space direction (i = 1, 2, (3))
j	Space direction (j = 1, 2, (3))
k	(Iteration) step number
	Space direction (k = 1, 2, (3))
m	Mean value
n	Time step number
	Normal
p	Fluid stress
	Plastic
q	Specific discharge
s	Solid
	Shear
t	Tangential
u	(Solid) displacement
+	(Positive) direction of Propagation
-	(Negative) direction of propagation
η	Related to non-linear model

Other symbols

$\cdot$	Superimposed dot: (first) time derivative
$\cdot\cdot$	Superimposed double dot: second time derivative
^	Amplitude of variable
	Numerical approximation
Δ	Laplace operator

APPENDIX 2A
NOTE ON THE FLOW ALONG A STRAIGHT INTERFACE

In this appendix it is shown that in a dynamical system, the relation between the discharge parallel an interface depends on the relative acceleration of the fluid.

The x-axis is chosen along the interface. The equation of motion of the fluid in the materials along the interface are (cf. Equation 2.17 in chapter 2)

$$n^a \frac{\partial \pi^a}{\partial x} - \frac{b^a}{n^a} q_x^a = n^a \rho_f \left(\ddot{u}_x^a + \frac{1}{n^a} \dot{q}_x^a\right) + \frac{\rho_a^a}{n^a} \dot{q}_x^a$$
$$n^b \frac{\partial \pi^b}{\partial x} - \frac{b^b}{n^b} q_x^b = n^b \rho_f \left(\ddot{u}_x^b + \frac{1}{n^b} \dot{q}_x^b\right) + \frac{\rho_a^b}{n^b} \dot{q}_x^b \qquad (2A.1)$$

with $b = n^2\mu/K$.

At the interface, no additional flow resistance is assumed ($\kappa_s = \infty$), and therefore $\pi^a = \pi^b$. Moreover, no slip is supposed: $u_x^b = u_x^a$. Substitution of these equalities in Equation (2A.1) and subtracting the resulting equations, one finds

$$\frac{\mu}{K^b} q_x^b - \frac{\mu}{K^a} q_x^a = \frac{1}{n^a}\left(\rho_f + \frac{\rho_a^a}{n^a}\right) \dot{q}_x^a - \frac{1}{n^b}\left(\rho_f + \frac{\rho_a^b}{n^b}\right) \dot{q}_x^b \qquad (2A.2)$$

The relation between the parallel discharges along the interface which is well known from stationary groundwater flow, reads

$$q_x^a = \frac{K^a}{K^b} q_x^b \Rightarrow \dot{q}_x^a = \frac{K^a}{K^b} \dot{q}_x^b \qquad (2A.3)$$

Introduction of this relation shows

$$\left[\frac{1}{n^a}\left(\rho_f + \frac{\rho_a^a}{n^a}\right) \frac{K^a}{K^b} - \frac{1}{n^b}\left(\rho_f + \frac{\rho_a^b}{n^b}\right)\right] \dot{q}_x^b = 0 \qquad (2A.4)$$

This equation holds if $\dot{q}_x^b = 0$ and thus $\dot{q}_x^a = 0$, this refers to the stationary case of groundwater flow, or if the material properties are as such that the first term of the left-hand-side is zero.

This shows that the general statical relation, shown in Equation (2A.3), does not always apply in dynamics of saturated porous media.

APPENDIX 4A
RAVIART-THOMAS INTERPOLATION ON TRIANGULAR ELEMENTS

A triangular element in x-y space is considered, see Figure 4A.1.

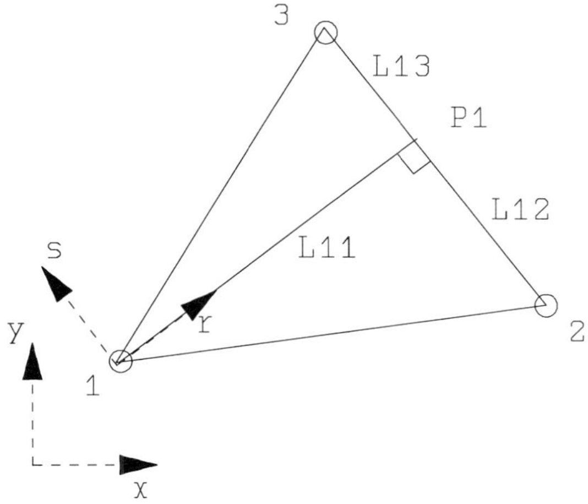

Figure 4A.1 Element and notation for Raviart-Thomas interpolation

The interpolation functions for edge 1 (opposite to node 1) are calculated in this appendix. The interpolation functions for the other two edges can be obtained by cyclic replacement of indices.

Firstly the point $P_1 = (X_1, Y_1)$ is calculated. The line between this point and node 1 (P_1-N_1) is perpendicular to the line between the nodes 2 and 3 (N_2-N_3).

$$X_1 = \frac{x_1(x_3-x_2)^2 + (y_3-y_2)[x_3(y_1-y_2)+x_2(y_3-y_1)]}{(x_3+x_2)^2+(y_3+y_2)^2}$$

$$Y_2 = \frac{y_1(y_3-y_2)^2 + (x_3-x_2)[y_3(x_1-x_1)+y_2(x_3-x_1)]}{(x_3+x_2)^2+(y_3+y_2)^2} \qquad (4A.1)$$

Three distances from the point P_1 to the three nodes are calculated

$$l_{1j} = \sqrt{(X_1-x_j)^2+(Y_1-y_j)^2} \qquad (4A.2)$$

with j = 1, 2 and 3.

An auxiliary coordinate-system (r,s) is defined in the element, see Figure 4A.1. The component of the discharge at edge 1 in r-direction q_r is constant along the edge

$$q_r = Q_1 \tag{4A.3}$$

The direction of the discharge at a location on edge 1 points through node 1, see Figure 4A.2. This is needed, since a value of Q_1 along edge 1 may not lead to a discharge through another edge. The discharge is chosen linear dependant on the coordinate r. Now the complete discharge in the element related to the discharge through edge 1 can be calculated

$$\begin{aligned} q_r &= \frac{r}{l_{11}} Q_1 \\ q_s &= \frac{s\,r}{l_{11}\,(l_{12}+l_{13})} Q_1 \end{aligned} \tag{4A.4}$$

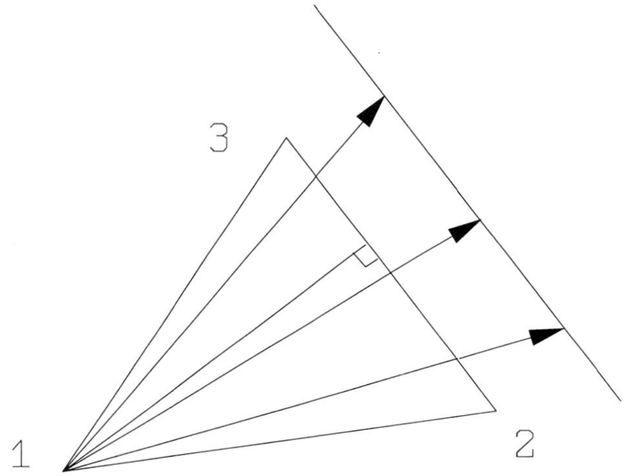

Figure 4A.2 Direction of discharge through edge 1

The discharge at the three nodes are calculated from this equations

$$\begin{aligned} q_r(x_1,y_1) &= 0 \qquad & q_s(x_1,y_1) &= 0 \\ q_r(x_2,y_2) &= Q_1 \qquad & q_s(x_2,y_2) &= Q_1 \frac{l_{12}}{l_{12}+l_{13}} \\ q_r(x_3,y_3) &= Q_1 \qquad & q_s(x_3,y_3) &= Q_1 \frac{l_{13}}{l_{12}+l_{13}} \end{aligned} \tag{4A.5}$$

The axis transformation from (r,s) to (x,y) is calculated from

$$q_x = q_r \cos\alpha - q_s \sin\alpha$$
$$q_y = q_r \sin\alpha + q_s \cos\alpha$$
with
$$\sin\alpha = \frac{(x_2 - x_3)}{l_{12} + l_{13}} \qquad (4A.6)$$
$$\cos\alpha = \frac{(y_2 - y_3)}{l_{12} + l_{13}}$$

α is the angle between the x-axis and the r-axis.

Finally, the interpolations are written in linear functions of the global coordinates x and y

$$q_x(x,y) = A_x + B_x x + C_x y$$
$$q_y(x,y) = A_y + B_y x + C_y y \qquad (4A.7)$$

The coefficients of these interpolations can be calculated from the discharges in the nodes. This leads to the following equations for A_x, B_x and C_x

$$q_x(x_1,y_1) = A_x + B_x x_1 + C_x y_1 = 0$$
$$q_x(x_2,y_2) = A_x + B_x x_2 + C_x y_2 = Q_1 [\cos\alpha - l_{12} \sin\alpha] \qquad (4A.8)$$
$$q_x(x_3,y_3) = X_x + B_x x_3 + C_x y_3 = Q_1 [\cos\alpha - l_{13} \sin\alpha]$$

thus

$$\begin{pmatrix} 1 & x_1 & y_1 \\ 1 & x_2 & y_2 \\ 1 & x_3 & y_3 \end{pmatrix} = \begin{pmatrix} A_x \\ B_x \\ C_x \end{pmatrix} = Q_1 \begin{pmatrix} 0 \\ \cos\alpha - l_{12} \sin\alpha \\ \cos\alpha - l_{13} \sin\alpha \end{pmatrix} = \begin{pmatrix} 0 \\ Q_{x2} \\ Q_{x3} \end{pmatrix} \qquad (4A.9)$$

The solution of this equation is

$$A_x = \frac{1}{D} [-Q_{x2}(x_1 y_3 - x_3 y_1) + Q_{x3}(x_1 y_2 - x_2 y_1)]$$
$$B_x = \frac{1}{D} [Q_{x2}(y_3 - y_1) - Q_{x3}(y_2 - y_1)] \qquad (4A.10)$$
$$C_x = \frac{1}{D} [-Q_{x2}(x_3 - x_1) + Q_{x3}(x_3 - x_1)]$$

with

$$D = (x_2 y_3 - x_3 y_2) - (x_1 y_3 - x_3 y_1) + (x_1 y_2 - x_2 y_1) \qquad (4A.11)$$

For the interpolation for q_y the equation for solving A_y, B_y and C_y is only slightly different from Equation (4A.9), (4A.12)

$$\begin{pmatrix} 1 & x_1 & y_1 \\ 1 & x_2 & y_2 \\ 1 & x_3 & y_3 \end{pmatrix} = \begin{pmatrix} A_y \\ B_y \\ C_y \end{pmatrix} = Q_1 \begin{pmatrix} 0 \\ (\sin\alpha + l_{12}\cos\alpha)\dfrac{l_{12}}{l_{12}+l_{13}} \\ (\sin\alpha + l_{13}\cos\alpha)\dfrac{l_{13}}{l_{12}+l_{13}} \end{pmatrix} = \begin{pmatrix} 0 \\ Q_{y2} \\ Q_{y3} \end{pmatrix} \tag{4A.12}$$

The solution is

$$\begin{aligned} A_y &= \frac{1}{D}\left[-Q_{y2}(x_1y_3 - x_3y_1) + Q_{y3}(x_1y_2 - x_2y_1)\right] \\ B_y &= \frac{1}{D}\left[Q_{y2}(y_3 - y_1) - Q_{y3}(y_2 - y_1)\right] \\ C_y &= \frac{1}{D}\left[-Q_{y2}(x_3 - x_1) + Q_{y3}(x_3 - x_1)\right] \end{aligned} \tag{4A.13}$$

with D defined in equation (4A.11).

Example

Now, the standard element is considered with $(x_1,y_1) = (0,0)$, $(x_2,y_2) = (1,0)$ and $(x_3,y_3) = (0,1)$. The solution $Q_1 = 1/2\ \sqrt{2}$, $Q_2 = 0$ and $Q_3 = -1$ is shown graphicly in Figure 4A.3. A constant discharge in r-direction is revealed.

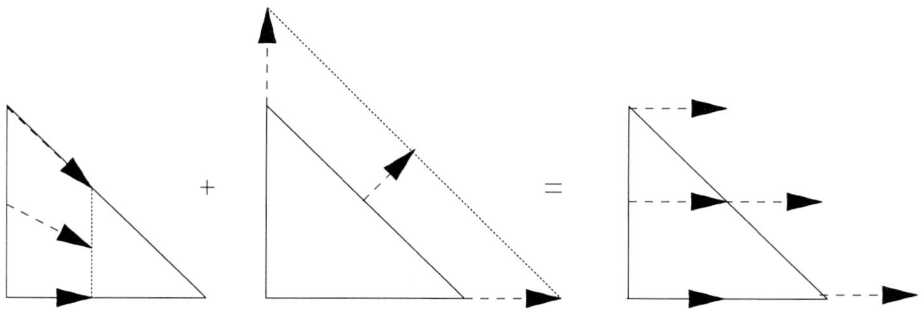

Figure 4A.3 Special case on standard element

APPENDIX 5A
DESIGN OF PILEDRIVING SIMULATOR

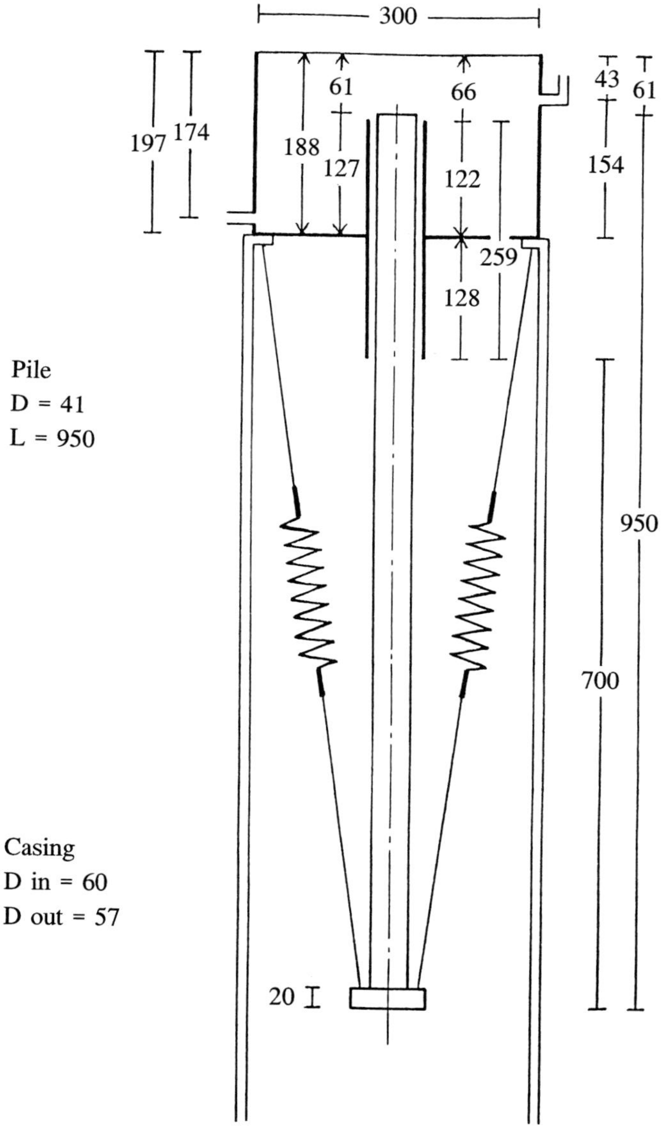

Figure 5A.1 Design of piledriving simulator measures in mm

APPENDIX 5B
THE MOTION OF THE SURFACE AND THE MOTION OF THE SAND RIDGE

Introduction

The motion of the surface and the motion of the sand ridge are correlated. In this appendix both motions are described and the correlation is shown. The comparison of these motions shows that the motion of the sand ridge is much slower than the motion of the surface of the sample. Since both motions are measured with different techniques, the applicability of these techniques can be checked with experimental data. Based on this evaluation it is concluded that the motion of the soil surface cannot be measured with the displacement transducers, due to the high frequency of this motion.

Models

Motion of the surface

The motion of the surface is described by a sinusoidal motion, positive in upward direction

$$u(t) = 0.5\, u_o [1 - \cos(\omega t)] \tag{5B.1}$$

with t the time, $0 < t < T$
T the duration of the motion
$\omega = 2\pi/T$ the angular frequency
u_o the maximum surface displacement

The velocity and the acceleration can be found by derivation with respect to time

$$\begin{aligned} v(t) &= 0.5\, u_o\, \omega \sin(\omega t) \\ a(t) &= 0.5\, u_o\, \omega^2 \cos(\omega t) \end{aligned} \tag{5B.2}$$

From the motion of the surface the initial velocity of the sand ridge can be calculated, since the free motion of the dry sand ridge starts at the moment that the acceleration of the soil surface is smaller than the acceleration due to gravity ($g = 9.81$ m/s^2). This is reached at moment $t = t_{los}$.

It is supposed that the maximum acceleration of the surface is much greater than the acceleration due to gravity.

$$a_{max} = \frac{1}{2} u_o \omega^2 \gg g \tag{5B.3}$$

Than the time t_{los} can be approximated by

$$a(t_{los}) = \frac{1}{2} u_o \omega^2 \cos(\omega t_{los}) = -g \approx 0 \tag{5B.4}$$

so

$$t_{los} = \frac{\pi}{2\omega} = \frac{1}{4} T \tag{5B.5}$$

The velocity of the surface at moment t_{los} (and thus the initial velocity of the dry sand ridge v_o) is found to be

$$v_o = v(t_{los}) = 0.5\, u_o\, \omega \tag{5B.6}$$

<u>Motion of sand ridge</u>

The sand ridge is considered as a free moving mass with initial velocity v_o at which the acceleration due to gravity works. Air friction is neglected (which is allowed since the measured curves of jump height depending on time are reasonable parabolas). The motion of the sand can be calculated as follows

$$s_m = v_o t_m - \frac{1}{2} g\, t_m^2 \tag{5B.7}$$

$$v_m = v_o - g t_m \tag{5B.8}$$

with
- s_m maximum height
- t_m time at which maximum height is reached
- v_o initial velocity
- g acceleration due to gravity (9.81 m/s^2)
- v_m velocity at maximum height (= 0.0 m/s)

Experimental data

The displacements of the surface are measured directly with two displacement transducers (Fairlight Optimess). Figure 5B.1 shows a result of one measurement. It shows an almost linear motion up to a displacement of 1 mm during 8 ms, followed by an almost linear motion back during 20 ms. From Figure 5B.1 one sees that the displacement of the surface is about 1.2 mm and the duration of the motion is about 30 ms.

From the high speed film it is derived that the jump height of the dry sand ridge is about 30 mm and the duration of the jump is about 0.15 s (=12/80 s: 12 s slow motion time divided by 80 as relative speed. The high speed film ran on 2000 pic/s and normal film on 25 pic/s).

Compared with the motion of the thin sand ridge, the motion of the surface has a much smaller magnitude (1.2 mm to 30 mm) and a shorter duration (30 ms to 150 ms).

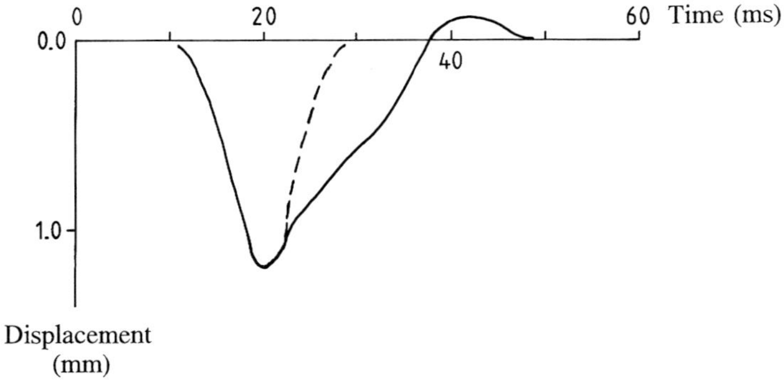

Figure 5B.1 Directly measured displacement of the surface

Model application

The value of u_o and T are estimated from the displacement measurement (see Figure 5B.1)

u_o = 1.2 mm = $1.2*10^{-3}$ m
T = 20 ms = $20*10^{-3}$ s
$v(t_{los}) = 0.5\ u_o\ \omega = \pi\ u_o / T = 0.18$ m/s

From the high speed film the value of the maximum height (s_m) is estimated about 30 mm. By using the Equations (5B.7) and (5B.8) the time with maximum velocity is approximated by

$$t_m = \sqrt{\frac{2s_m}{g}} = \sqrt{\frac{2*30*10^{-3}}{10}} = 0.08s \qquad (5B.9)$$

This time t_m is half the duration of the jump. The initial velocity is $v_o = g\ t_m = 0.8$ m/s.

The initial velocity based on the high speed film (0.8 m/s) differs a factor 4 from the value based on the displacement measurement (0.18 mm). The difference may be explained by the fact that the displacement transducers are loaded faster than their maximum speed. According to the product information the band width is 200 Hz.

Let the lowest frequency in the surface motion be 200 Hz (T = 5 ms). Then the initial velocity of the dry sand ridge is estimated by

$v(t_{los}) = 0.5\ u_o\ \omega = \pi\ u_o / T = 0.75$ m/s

This value fits better to the results based on the high speed film. It means that the lowest frequency in the motion of the soil surface equals the highest frequency in the displacement transducer. The displacement transducers are not suitable for application in this experiment.

The maximum acceleration must still be compared with the acceleration due to gravity

$$a_{max} = \frac{1}{2} u_o \omega^2 = \frac{2\pi^2 u_o}{T^2} \qquad (5B.10)$$

For T = 20 ms one finds a_{max} = 59 m/s^2 = 6g and for T = 5 ms on finds a_{max} = 950 m/s^2 = 95g. For the lower frequency the approximation is questionable, for the higher frequency the approximation is justified.

Conclusion

The values of the two measurement techniques are not in accordance. Using a simple description of the motion, the difference can be explained by the fact that the fast motion of the surface is not measured accurately by the displacement transducers. This justifies the indirect method of observation by using the sand ridge.

APPENDIX 5C
DERIVATION OF RELATION BETWEEN STRESS DEVIATOR AND INTENSITY OF THE LIGHT

In this appendix the relation between the stress in the sample and the light intensity is derived quantitatively. The derivation is shown for one wavelength only. It holds for all wavelengths if all components are insensitive for the wavelength. For the quarter lambda plates this insensitivity for wavelength is questionable.

The outline of the setup is shown in Figure 5C.1. The light source is at the left hand side. The camera at the right hand side. The coordinate systems used in the derivation are shown in Figure 5C.2.

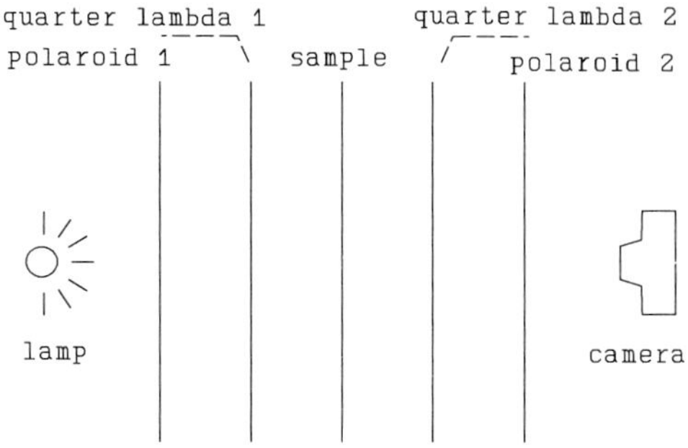

Figure 5C.1 Outline of test set-up

The light source generates unpolarised light. After passing the first polaroid filter the light can be described by

$$\begin{aligned} q_x &= 0 \\ q_y &= A \sin(\omega\, t) \end{aligned} \tag{5C.1}$$

with:

	q_i	motion in direction i
	A	initial amplitude
	ω	(angular) frequency of the light
	t	time

The first quarter lambda plate creates circular polarized light from the polarised light. Before the quarter lambda plate the components of the light are (in the local coordinate system for the quarter lambda plate)

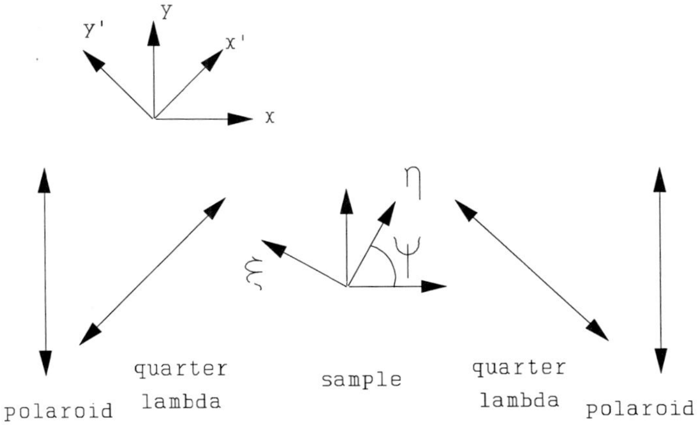

Figure 5C.2 Coordinate systems used in derivation

$$q_{x'} = \frac{1}{2} A \sqrt{2} \sin(\omega t)$$
$$q_{y'} = \frac{1}{2} A \sqrt{2} \sin(\omega t) \tag{5C.2}$$

After the quarter lambda plate the components of the light are (in the local coordinate system for the quarter lambda plate)

$$q_{x'} = \frac{1}{2} A \sqrt{2} \sin(\omega t - \frac{1}{2}\pi) = -\frac{1}{2} A \sqrt{2} \cos(\omega t)$$
$$q_{y'} = \frac{1}{2} A \sqrt{2} \sin(\omega t) \tag{5C.3}$$

The result is described by

$$q_x = -\frac{1}{2} A \cos(\omega t) - \frac{1}{2} A \sin(\omega t)$$
$$q_y = -\frac{1}{2} A \cos(\omega t) + \frac{1}{2} A \sin(\omega t) \tag{5C.4}$$

in the original coordinate system.

Now, the light is transmitted through the sample. In the sample, the internal stresses lead to the mean stresses σ_1 and σ_2. The first mean stress make an angle ψ with the horizontal axis (x) and an angle $\phi = \psi + \pi/2$ with the axis x'. Due to the stress in the sample, a retardation δ occurs in the η-direction. This retardation δ is defined by [cf. Allersma, 1987]

$$\delta = M(\sigma_1 - \sigma_2) \tag{5C.5}$$

with M the photo-elastic constant of the material.

Before the sample, the light is decomposed in the η,ξ-system by

$$q_\eta = -\frac{1}{2}\sqrt{2}\ A\ \cos(\omega t + \phi)$$
$$q_\xi = \frac{1}{2}\sqrt{2}\ A\ \sin(\omega t + \phi) \qquad (5C.6)$$

The sample retards the light in η-direction with a retardation of δ, thus the decomposition in the η, ξ-system are

$$q_\eta = -\frac{1}{2}\sqrt{2}\ A\ \cos(\omega t + \phi + \delta)$$
$$q_\xi = \frac{1}{2}\sqrt{2}\ A\ \sin(\omega t + \phi) \qquad (5C.7)$$

These equations are rewritten to the x', y'-system

$$q_{x'} = -\frac{1}{2}\sqrt{2}\ A\ \cos(\omega t + \phi + \delta)\cos\phi - \frac{1}{2}\sqrt{2}\ A\ \sin(\omega t + \phi)\sin\phi$$
$$q_{y'} = -\frac{1}{2}\sqrt{2}\ A\ \cos(\omega t + \phi + \delta)\cos\phi + \frac{1}{2}\sqrt{2}\ A\ \sin(\omega t + \phi)\cos\phi \qquad (5C.8)$$

The quarter lambda plate retards the light in y'-direction with a retardation of π/2 thus

$$q_{x'} = -\frac{1}{2}\sqrt{2}\ A\ \cos(\omega t + \phi + \delta)\cos\phi - \frac{1}{2}\sqrt{2}\ A\ \sin(\omega t + \phi)\sin\phi$$
$$q_{y'} = -\frac{1}{2}\sqrt{2}\ A\ \cos(\omega t + \phi + \delta - \frac{1}{2}\pi)\cos\phi + \frac{1}{2}\sqrt{2}\ A\ \sin(\omega t + \phi - \frac{1}{2}\pi)\cos\phi \qquad (5C.9)$$

Finally, the component in the y-direction is removed by the second quarter lambda plate. The result is

$$q_x = -\frac{1}{2}A\ [\ \cos(\omega t + 2\phi)\cos\delta - \sin(\omega t + 2\phi)\sin\delta - \cos(\omega t + 2\phi)\ ] \qquad (5C.10)$$
$$\textit{and} \quad q_y = 0$$

This equation gives the light behind the test set-up. For the special case $\delta = 0$, q_x is zero. This is the case when there is no sample in the test set-up or the two mean stresses are equal. It is called the dark field situation.

Finally, the light intensity can be rewritten to

$$q_x = \frac{1}{2}\sqrt{2}\ A\sqrt{1-\cos\delta}\ \sin(\omega t + 2\phi + \gamma) \qquad (5C.11)$$
$$q_y = 0$$

The additional phase angle γ still depends on the retardation angle δ

$$\tan\gamma = \frac{1-\cos\delta}{\sin\delta} \qquad (5C.12)$$

The values of the amplitude $B = \frac{1}{2}\sqrt{2}\ A\ \sqrt{(1-\cos\delta)}$ are shown in Figure 5C.3 for several values of the retardation angle δ. For very small values of δ, the relation is linear. However, the relation is periodic in δ with period 2π.

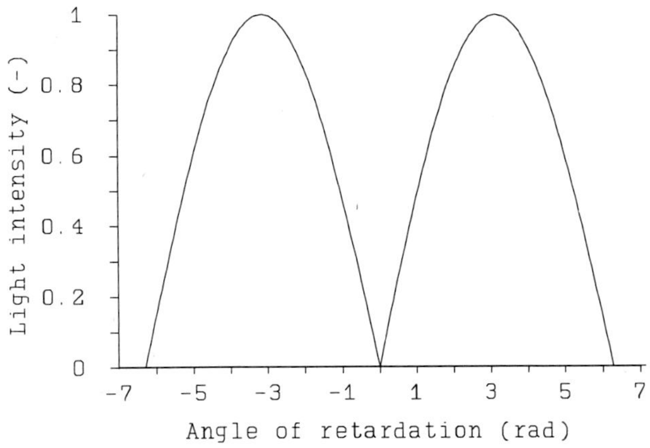

Figure 5C.3 Light-intensity depending on retardation angle

APPENDIX 5D
MECHANICAL PROPERTIES OF GLASS-OIL MIXTURE

Table 5D.1 Mechanical properties of glass-oil mixture

	name		unit	value	source
K_f	bulk modulus oil		MPa	1920	measured
ρ_f	volumetric mass oil		kg/m^3	875	Shell [undated]
ν	kinematic viscosity (20°C)		mm^2/s	37	Shell [undated]
ρ_s	volumetric mass glass		kg/m^3	2230	Allersma [1987]
n	porosity	loose dense	-	0.52 0.43	measured
K	intrinsic permeability	loose dense	mm^2	0.034 0.013	measured
E_s	Young's modulus glass		MPa	45	estimated from Allersma [1987]
ν_s	Poison ratio glass		-	0.33	assumed
E_{steel}	Young's modulus steel		MPa	210 000	
ν_{steel}	Poison ratio steel		-	0.3	
ρ_{steel}	volumetric mass steel		kg/m^3	7850	

APPENDIX 6A
CALCULATION OF THE FORCE AT THE PILE HEAD

Introduction

The hammer blow by the ball hitting the cushion is modelled in the finite element calculation by a force on the pile head. In this appendix a simple discrete model is used in order to derive the force on the pile head. The pile is assumed to be long. Therefore, the influence of wave reflections from the pile toe are neglected.

Simple model

The ball-cushion-pile system is considered as a mass on a spring which is supported by a dashpot. The mass hits the cushion with an initial velocity v_0. Figure 6A.1 shows the problem schematisation. The following values are adopted:

mass	$m = 0.122$ kg
dashpot	$c = 1261$ Ns/m
spring	$k = 4.0*10^6$ N/m
velocity	$v_0 = 5.7$ m/s

The influence of the acceleration due to gravity is neglected.

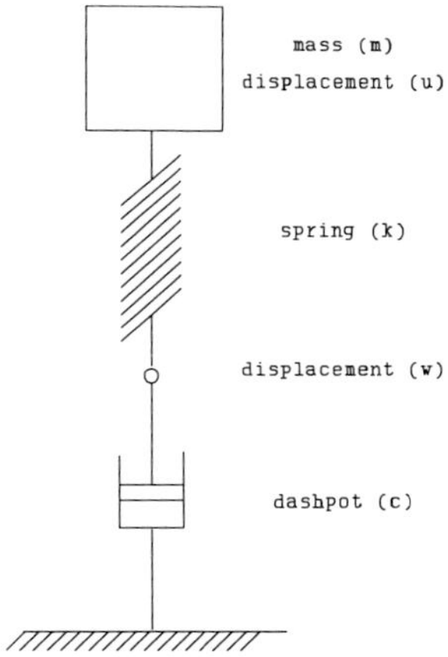

Figure 6A.1 Problem schematisation

General solution

The equations of motion are

$$m \frac{d^2u}{dt^2} + k\,u - k\,w = 0$$
$$c \frac{dw}{dt} - ku + kw = 0 \qquad (6A.1)$$

with u displacement of mass
w displacement of pile head

By introducing the velocity v of the mass as additional variable (v = du/dt), the equations of motion are written as a system of first order differential equations

$$\begin{pmatrix} 1 & 0 & 0 \\ 0 & m & 0 \\ 0 & 0 & c \end{pmatrix} \frac{d}{dt} \begin{pmatrix} u \\ v \\ w \end{pmatrix} + \begin{pmatrix} 0 & -1 & 0 \\ k & 0 & -k \\ -k & 0 & k \end{pmatrix} \begin{pmatrix} u \\ v \\ w \end{pmatrix} = \begin{pmatrix} 0 \\ 0 \\ 0 \end{pmatrix} \tag{6A.2}$$

The solution of this is system is found by testing an $e^{\beta t}$ solution, and solving β for non-trivial solutions. This leads to three values of β, the eigen values,

$$\begin{aligned} \beta &= 0 \\ \beta &= -\alpha + i\omega \\ \beta &= -\alpha - i\omega \end{aligned} \tag{6A.3}$$

$$\textit{with} \quad \alpha = \frac{k}{2c} \quad \textit{and} \quad \omega = \frac{k}{2c}\sqrt{\frac{4c^2}{mk} - 1}$$

Based on the values for the mass, dashpot constant and spring constant, the value of ω is real. A general solution of the equations of motion is found by calculating the eigen vectors for each eigen value β, showing

$$\begin{pmatrix} u \\ v \\ w \end{pmatrix} = A_0 \begin{pmatrix} 1 \\ 0 \\ 1 \end{pmatrix} + A_1 \begin{pmatrix} 1 \\ \alpha + i\omega \\ \dfrac{2\alpha}{3\alpha + i\omega} \end{pmatrix} e^{-\alpha t} e^{i\omega t} + A_2 \begin{pmatrix} 1 \\ (\alpha - i\omega) \\ \dfrac{2\alpha}{3\alpha - i\omega} \end{pmatrix} e^{-\alpha t} e^{-i\omega t} \tag{6A.4}$$

Finally the initial conditions at t = 0 are introduced (u(0) = 0, v(0) = v_0 and w(0) = 0). This shows

$$\begin{aligned} A_0 &= \frac{2\alpha}{\alpha^2 + \omega^2} v_0 \\ A_1 &= \frac{(\alpha - i\omega)(3\alpha + i\omega)}{2i\omega(\alpha^2 + \omega^2)} v_0 \\ A_2 &= -\frac{(\alpha + i\omega)(3\alpha - i\omega)}{2i\omega(\alpha^2 + \omega^2)} v_0 \end{aligned} \tag{6A.5}$$

Motion of pile head

Now, the motion of the pile head (w) is studied. The solution reads

$$w(t) = \frac{2\alpha v_0}{\alpha^2 + \omega^2} \left[1 + \frac{\alpha - i\omega}{2i\omega} e^{-\alpha t} e^{i\omega t} - \frac{\alpha + i\omega}{2i\omega} e^{-\alpha t} e^{-i\omega t} \right] \tag{6A.6}$$

This leads to a real expression for the displacement of the pile head

$$w(t) = \frac{2\alpha v_0}{\alpha^2 + \omega^2} \left(1 - [\cos(\omega t) - \frac{\alpha}{\omega} \sin(\omega t)]\, e^{-\alpha t} \right) \tag{6A.7}$$

Derivation with respect to time gives the velocity of the pile head

$$\frac{dw}{dt} = \frac{2\alpha v_0}{\alpha^2 + \omega^2}\, e^{-\alpha t} \left(2\alpha\cos(\omega t) + \frac{\omega^2 - \alpha^2}{\omega} \sin(\omega t) \right) \tag{6A.8}$$

After multiplication of the velocity with the pile impedance (c), the force on the pile head is found.

Application of the model

For several values of the stiffness of the cushion the velocity of the pile head is shown in Figure 6A.2. The maximum velocity occurs at t ≈ $0.2*10^{-3}$ s. The maximum value is 2.8 m/s. At t = $0.7*10^{-3}$ s the velocity of the pile head changes of sign. Therefore, tension occurs and the connection between the ball and the pile will be broken.
The wave velocity in the pile is 1 m/ms. During the loading time, the wave has propagated 0.8 m, which is less than two times the length of the pile.

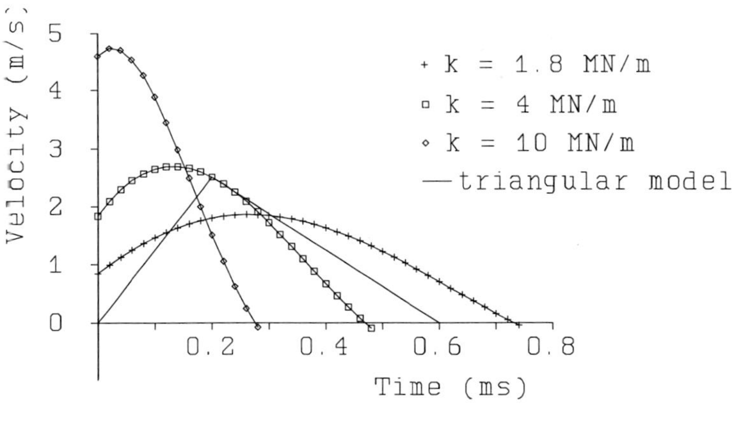

Figure 6A.2 Dependency of velocity pile head on stiffness of cushion

CURRICULUM VITAE

Naam

Paul Hölscher

Geboren

23 februari 1960 te Velsen

Opleiding

1972-1978	Gymnasium β te Velsen
1978-1985	Technische Universiteit Delft, afdeling der Civiele Techniek, Afstudeerrichting Grondslagen der Civiele Techniek, ingenieursdiploma met lof

Loopbaan

1984	Stage bij het National Road Laboratory te Roskilde, Road Directorate, Ministry of Transport, Denemarken
1985-heden	Projectleider bij Grondmechanica Delft, afdeling Wiskunde en Informatica Gedurende de jaren 1990-1994 is het promotie onderzoek uitgevoerd aan de Technische Universiteit Delft, afdeling der Civiele Techniek, vakgroep Waterbouwkunde, sectie Geotechniek.